TIME AND MEMORY

Time and Memory is one of a series of low-cost books under the title PSYCHOANALYTIC **ideas** which brings together the best of Public Lectures and other writings given by analysts of the British Psychoanalytical Society on important psychoanalytic subjects.

The books can be ordered from:
Karnac Books
www.karnacbooks.com
Tel. +(0)20 7431 1075
Fax: +(0)20 7435 9076
E-mail: shop@karnacbooks.com

Other titles in the Psychoanalytic ideas Series:

Shame and Jealousy: The Hidden Turmoils
Phil Mollon

Dreaming and Thinking
Rosine Jozef Perelberg

Spilt Milk: Perinatal Loss and Breakdown
Joan Raphael-Leff (editor)

Unconscious Phantasy
Riccardo Steiner (editor)

Psychosis (Madness)
Paul Williams (editor)

Adolescence
Inge Wise (editor)

Child Analysis Today
Luis Rodríguez de la Sierra (editor)

Psychoanalytical Ideas and Shakespeare
Inge Wise and Maggie Mills (editors)

You Ought To!
Bernard Barnett

Symbolization
Jim Rose (editor)

TIME AND MEMORY

Editor

Rosine Jozef Perelberg

Series Editors

Inge Wise and *Paul Williams*

KARNAC

First published in 2007 by the Institute of Psycho-Analysis, London
Karnac Books Ltd
118 Finchley Road
London NW3 5HT

British Library Cataloguing in Publication Data

A C.I.P. for this book is available from the British Library

ISBN 978 1 85575 434 8

Designed typeset and produced by The Studio Publishing Services Ltd
www.publishingservicesuk.co.uk
e-mail: studio@publishingservicesuk.co.uk

Printed in Great Britain

10 9 8 7 6 5 4 3 2 1

www.karnacbooks.com

CONTENTS

ACKNOWLEDGEMENTS

I would like to thank Paul Williams and Inge Wise, the editors of the Psychoanalytic Ideas series, for their support in the production of this book, and all the contributors for their commitment to this project.

My thanks also go to *The International Journal of Psychoanalysis* for permission to reproduce the following papers:

- Bronstein, C. (2002). Borges, immortality and the circular ruins. *International Journal Psychoanalysis, 83*: 647–660.
- Rose, J. (1997). Distortions of time in the transference: some clinical and theoretical implication. *International Journal Psychoanalysis, 78*: 453–468.

A number of the chapters in this book were originally published elsewhere. So I am grateful to:

- Free Associations for permission to reproduce André Green's chapter, originally published in *Time in Psychoanalysis*, London: Free Associations (pp. 9–27).
- Karnac for permission to reproduce Gregorio Kohon's chapter, originally published in *No Lost Certainties to be Recovered,*

London: Karnac (pp. 123–146), here published with some revisions.

- *The Psychoanalytic Quarterly* for permission to reproduce David Bell's paper, "Existence in time: development or catastrophe?", *Psychoanalytic Quarterly, 75*: 783–805.

ABOUT THE CONTRIBUTORS

David Bell is a training and supervising psychoanalyst of the British Psychoanalytic Society, where he is also Chairman of the Scientific Committee. He is a Consultant Psychiatrist in Psychotherapy at the Tavistock Clinic, where he lectures on Psychoanalytic Theory and is Director of a Specialist Unit for severe psychological disturbance. He lectures and publishes widely and has edited *Psychoanalysis and Culture: A Kleinian Perspective,* and *Reason and Passion: The Work of Hannah Segal.*

Catalina Bronstein (nee Halperin) is a training analyst of the British Institute of Psychoanalysis. She studied medicine and qualified as a psychiatrist in Buenos Aires, Argentina. She trained as a child psychotherapist at the Tavistock Clinic, as a psychoanalyst at the British Institute of Psychoanalysis, and now works at the Brent Adolescent Centre and in private practice. She is Senior Lecturer (Honorary) in Psychoanalytic Theory at University College London, is UK Editor of the *International Journal of Psychoanalysis* and Associate Editor of the New Library of Psychoanalysis. She edited *Kleinian Theory. A Contemporary Perspective* (2001).

Rosemary Davies is a training and supervising analyst of the British Psychoanalytical Society. She teaches and lectures in the UK and abroad and is Chair of the Curriculum of the Masters programme in psychoanalytic theory at University College London. She has edited and contributed to various psychoanalytic books and more recently to the "The analyst at work" series in the *International Journal of Psychoanalysis*. She works in private practice and supervises in the National Health Service in London.

André Green is an Honorary Member of Paris Psychoanalytical Society, an Honorary Member of British Psychoanalytical Society, ex-Professor Freud Memorial Chair, and Honorary Professor Buenos Aires University. He is the author of over twenty books, among which are: *On Private Madness* (1986), *The Work of the Negative* (1999), *Life Narcissism and Death Narcissism* (2001).

Gregorio Kohon is a training analyst of the British Psychoanalytical Society. He edited *The British School of Psychoanalysis—The Independent Tradition* (Free Association, 1986), and *The Dead Mother—The Work of André Green* (Routledge, 1999). He also published *No Lost Certainties to be Recovered* (Karnac, 1999) and *Love and its Vicissitudes* (co-authored with André Green) (Routledge, 2005). Kohon has also published three books of poetry in Spanish. His novel *Papagayo Rojo, Pata de Palo* (Finalist, 2001 Fernando Lara Prize, Editorial Planeta, Barcelona) was published by Libros del Zorzal (Buenos Aires, 2003. An English version of this book, *Red Parrot, Wooden Leg*, is due to be published by Karnac (2007). At present, he works in London in private practice.

Rosine Jozef Perelberg, PhD, is a training analyst and supervisor, and a Member of the British Psychoanalytical Society, where she is currently Chair of the Curriculum Committee. She is Honorary Senior Lecturer in Psychoanalytic Theory at University College, London. She co-edited with Joan Raphael-Leff *Female Experience: Three Generations of British Women Psychoanalysts on Work with Women* (1997). She has edited *Psychoanalytic Understanding of Violence and Suicide* (1998), *Dreaming and Thinking* (2000, 2003), and *Freud: A Modern Reader* (2005). Her book *Time, Space and Phantasy* will be published in 2008 in the New Library of Psychoanalysis.

James Rose, PhD, is a Fellow of the Institute of Psychoanalysis and a member of the British Psychoanalytical Society. He has a private psychoanalytic practice in London. He has worked in the Brandon Centre, an inner city charity specializing in the psychotherapeutic treatment of adolescents and young adults for the past twenty years.

Paul Williams is a Consultant Psychotherapist in the British National Health Service in Belfast, Northern Ireland; a Training and Supervising Analyst at the British Psychoanalytic Society; a Professor at Queens University Belfast; a Member of the Royal Anthropological Society; and former Joint Editor-in-Chief of the *International Journal of Psychoanalysis.*

Introduction

Rosine Jozef Perelberg

The origins of this book can be traced back to 1997, when I organized a day of Public Lectures on the theme of "Time and memory". Paul Williams' paper on "Making time: killing time" and David Bell's on "Existence in time: development or catastrophe?" were both presented on that occasion, together with my paper "To be or not to be—here".[1] André Green, James Rose, Gregorio Kohon, Rosemary Davies, and Catalina Bronstein were then invited to contribute to this collection.

In the first chapter, Green traces the various notion of time in Freud's work, from *Studies on Hysteria* to *Analysis Terminable and Interminable.* What comes to the fore is the complexity of Freud's views on temporality. Development, fixation, regression, repression, the return of the repressed, the timelessness of the unconscious, and *Nachträglichkeit* are some of the axes that permeate Freud's writings, in his discovery of a temporality that is truly psychoanalytical, and does not follow traditional, linear sequences of chronological time. Other psychoanalysts have since tended to select one or another aspect of Freud's complex notions of time and regard it as predominant, and have, in this way, selected to emphasize either a genetic or a structural perspective.

In *The Interpretation of Dreams*, Freud put forward the conception of the bi-directional nature of psychical processes. Progressive and regressive movements are in operation, as cathexes traverse the psychic space; the hypothesis of the timelessness of the unconscious is also explored in this work (see Chapter One). Dreams enabled Freud to discover a dismembered temporality, the rules of primary process, and the timelessness of the unconscious. In the *Three Essays on Sexuality*, the theory of infantile sexuality returned to a time that was ordered traditionally, that is, in terms of the growth characteristic of life. Green points out that if it is true that speaking of infantile sexuality was revolutionary in itself, the temporality that is emphasized in the *Three Essays* is a sequential, diachronic one, following the path of life.

Freud's work wavers between, on the one hand, a diachronic perspective grounded in childhood in which sexuality, unconscious desire, and, from 1918, an emphasis on the importance of the loss of the object are stressed, and on the other, a structural perspective of opposing psychical systems that are organized differently. Green suggests that in fact the two axes, historical and structural, are complementary.

With the development of the notion of primal phantasies, especially in *The Wolf Man*, a new temporal characteristic of the human psyche is introduced, which is, in Green's words the *disposition to re-acquisition*. Primal fantasies are re-actualized through individual experience. The diachronic heterogeneity of the psychical apparatus is accentuated, owing to the difference of structure between the agencies and the way in which the effects of the various forms of temporality are inscribed in them. Time is now not just in pieces; its parts are in a state of tension with each other.

It is also in the text on the Wolf Man that Freud more fully develops the notion of *Nachträglichkeit* (*après coup*), that refers to a theory of temporality that breaks with traditional notions of time as linear and sequential. Freud had already discussed this notion in the case of Emma (Freud & Breuer, 1895d). An earlier occurrence leaves its mark on the child, who cannot know the effect it will have on his subjectivity because he does not fully understand the way in which he has been affected by it. For this, the effects of post-pubertal sexuality at the two levels of accession to sexual maturity and intellectual development are necessary. In a recent paper (Perelberg, 2006),

I have indicated the network of concepts in which *Nachträglichkeit* is embedded, which includes the Freudian notions of infantile sexuality, trauma, castration, seduction, and primal scene. I have proposed the distinction between the *descriptive après coup* and the *dynamic après coup*. I have, furthermore, indicated that the dynamic *après coup* works as an illumination that gives colour to all other notions of time in Freud's work (Perelberg, 2006).

The various chapters of this book explore how the psychoanalytic notions of time can find expression in clinical practice, and shed light on historical events or literary creations.

The perception that patients in the analytic setting reproduce in the relationship with their analysts their internal experiences of time is discussed in the chapters by James Rose, David Bell, Paul Williams, and Catalina Bronstein. In the second chapter, James Rose suggests that patients create a characteristic sense of time and space in the transference. Patients' distortions of time reflect their psychopathology as well as their reactions to the temporal aspects of the psychoanalytic setting. Rose points out the distinction between phenomenological experience of time and chronological time. Patients who deny the passage of time are resistant to change, to mourning, and, ultimately, death. They also, at times, deny bodily and psychic development. Rose provides two case histories illustrating clinical fixation, in the first case, and psychic retreat in the second case. These examples are compared in order to demonstrate the unconscious processes underlying the particular time distortions being considered, their impact on the patients' lives, and their manifestation in the clinical setting. From these studies, Rose suggests that the asymmetry of the "arrow of time" cannot be assumed in the structure of psychic reality. Psychic reality is discontinuous and the structure of the discontinuities are revealed by the impact on the patient of the temporal aspects of the psychoanalytic setting.

In the third chapter, Paul Williams indicates that a psychoanalytic treatment gives access to the way in which differences in the experience of temporality are related to intrapsychic relations between the ego and its internal objects, in particular the capacity of the ego to tolerate separation and loss as durational phenomena, an essential aspect of development. Williams gives three examples of the experience of time from patients in psychoanalysis, each experience reflecting a different psychic state. As also suggested by

Green, Rose, and later Bell and Bronstein, Williams points out that making time is a depressive-position activity, necessitating acknowledgement of the other's separateness and significance to oneself and of one's role in the existence and well-being of the other person. To be unaware of time—to kill time—is to disavow the need for the other, and the other in oneself. In the example of a mildly agoraphobic patient in her thirties who is consistently late for sessions, Williams suggests that, by keeping the analyst waiting and finding him there, the patient experiences the reassuring feeling that the analyst does not intend to leave her.

The borderline personality kills time in order to keep apart from his objects. Williams acknowledges that there are expressions of killing time that are not necessarily symptoms of borderline states, but that also signify disavowal of object need. Compulsive daydreaming, dependency on alcohol or drugs, perversions, promiscuity, or living it up in the belief that this brings happiness—all these abolish a sense of time and dependency on others.

According to Williams, the most dramatic reaction *against* time occurs in psychosis. The psychotic patient, as Freud observed, detaches himself from external, temporal reality and lives in an internal, timeless reality. In the example that Williams presents, the patient felt disconnected, boundaryless, timeless, and engulfed—wholly persecuted by her projections. Her awareness of time had collapsed.

In the next chapter, Bell also emphasizes that viewing oneself as existing in time is an important developmental achievement. For some individuals, however, it is felt to be a permanent imminent catastrophe, evaded by the creation of a timeless world where, apparently, nothing ever changes; an illusion of time standing still. In his analysis of Oscar Wilde's *The Picture of Dorian Gray*, Bell suggests that the core of the narrative is the deep understanding the author brings to

> the cost to character of evasion of the facts of life, which include the inevitability of ageing and death, the feelings of guilt that are part of life and that lend it its moral force—all of which are supported by the awareness of the passage of time. [p. 71]

Dorian's solution to this painful conflict is perverse. Rather than adapting to reality, reality itself is altered; rather than bear the pain of the loss of his ideal self, he preserves it forever by exchanging places with the portrait—he will remain forever young, while the portrait will bear all the marks of time's passing.

Bell then gives an example of a patient who, like Dorian, lived in an illusory world and feared that any movement from this state would precipitate the breakdown that had brought her to analysis. She lived in a timeless world where there was no development, the only alternative being to feel precipitated into a terrifying world in which all the time evaded would suddenly catch up with her, leading to terrors of sudden disintegration and death. In his chapter, Bell also indicates that the capacity to mourn, and to bear guilt and loss, are essential to being able to fully understand oneself as existing in time.

Rosemary Davies' chapter makes a plea for a technique that recognizes the therapeutic value of regression, without ignoring the destructiveness inherent in the discovery of the object's otherness. She argues that in the psychoanalytic literature there remains a subtle manifestation of the distaste for the concept of regression in contemporary debate: a false dichotomy between those who argue that regression should be fostered and those who assert that it is tantamount to a psychic withdrawal. Davies takes the view that if, on the one hand, this state is a return to an earlier "frozen experience" (Winnicott, 1954a, p. 86 in this book), on the other hand, it may be an entirely new experience for the patient who has taken the risk of letting himself be known in this narcissistically vulnerable state. Where primary process thinking dominates, the potential for change may be greater, since defences are breached and access to unconscious material may be enhanced.

Davies suggests that the regressed state is ruptured by the recognition of dependence within the analysis: this can be a life or death moment, as hate finds expression when the patient emerges from the regression. Destructiveness is definitively present as the subject recognizes the presence of a discrete other, the "otherness" of the other (p. 92). The regressed patient requires the analyst to recognize when to hold off and when to hold forth. Recognizing the affective self faces the subject with the notion of the other: "The affect is the epiphany of the other for the subject" (Green, 1999, p. 215). The naming of affect may leave the patient feeling deficient and the patient's psychic survival may depend on the disavowal of the existence of another who may evoke intolerable feelings.

Davies views the analyst as a regulator of the regression "through the creative, receptive attitude, not through breaches of the setting". The analytic stance and the rules of abstinence are maintained, without forgoing a view that regression is a viable and interpretable state. The analytic stance is tested not only during the regression, but particularly as the patient discovers the hitherto "unknown environment", as Winnicott described it. Davies gives the clinical example of a patient who managed gradually to recover and recount something of his own history that had been lost to him. He was thus able to embark on what Loewald (1960) described as one of the aims of psychoanalysis: to restore the patient's sense of historicity "turning ghosts into ancestors" (Davies, p. 93).

In his chapter, Kohon utilizes the concept of *Nachträglichkeit*, in order to examine from a psychoanalytic perspective on temporality two historical events: the conquest of Mexico, and the siege of Masada. Freud always believed that every subject, as well as every nation, revises past events at a later date; this revision is what creates a historical past, imparting meaning to those events. Occurrences in the (mythical or historical) past, which could not be incorporated in a meaningful context at the time (thus, they were traumatic), are revised *nachträglich* so as to give significance to them *a posteriori* in the present. This creation of a meaningful past (what Freud (1986) called "a retranscription") helps to structure the invention of the present; that is, it determines the way we perceive the world and how we know it, the way we construct our knowledge of the present. This in turn might determine the construction of the future through the compulsion to repeat. "Nothing can be reduced to a linear sequence of events; logical time transforms the subject's past into a historicization of his present" (p. 106).

In relation to the conquest of Mexico, Kohon suggests that for the indigenous peoples of America, any event that had been experienced as strange or unexpected *before* the arrival of the Spaniards was then identified *retrospectively* as a message from the gods. The actions of humans were almost irrelevant to the outcome of a history that had already been written.

The Amerindians revised past events at a later date, through the process that Freud described as taking place *nachträlich*. The "omens" were given significance through the events that followed them. Pre-Columbian people made sense of the present in this way, and this made it easier for them to accept whatever events took place, since they had been "announced" in the past. The chronicles written by the Indians (whether Aztec, Mayan, or other) understood the defeat (the secondary event) through a retroactive reference to earlier "omens". It is only then that the primary trauma gained a new meaning, proving its psychical effectiveness:

> Masada, the fortress built at the top of a rocky mountain of the Judean desert, was the scene of a dramatic episode in the history of the Jewish people. It was the last stronghold ... held by the Jews against the Romans, three years after the Roman conquest of the rest of Judea and the destruction of the second Temple of Jerusalem by Titus in 70 AD. Between nine hundred and a thousand children, women, and men ... died at the top of this 1,300-foot mountain overlooking the Dead Sea, after a lengthy siege laid by the Romans. This episode that is perceived as the heroic sacrifice of the Jews to protect themselves against the invaders, is revisited in many instances of the history of Israel; representations of the past are used in terms of the needs of the present which retranslate the past. Memories are themselves interpretations. [p. 116]

In her chapter, Catalina Bronstein examines two stories by Borges. "The circular ruins", written in 1944, and "The immortal" (1949), from his book *The Aleph*.

"The circular ruins" is a short story about the struggle of an individual's determination to generate a man. An old man, a magician, gets off a boat, and installs himself in the ruins of an ancient circular temple. The Old Man is locked, encircled by, and inside the ruins of his own making. He cannot step outside his dream of dreaming a man, as he himself is the product of somebody else's dream. Bronstein understands the narrative as "a metaphor for the psychic struggle experienced by some individuals when they are confronted with feelings of pain and loss and their subsequent compulsive search for an omnipotent solution through the phantasy of becoming immortal" (p. 135).

Both stories deal with the omnipotent phantasy of cancelling chronological, linear time. Procreation can happen without sexual intercourse between a man and a woman. In "The immortal", Borges describes the result when he says, "There is nothing that is not as if lost in a maze of indefatigable mirrors".

Bronstein then discusses the clinical example of Mr A, a man who feels no desires but is aware of others who seem to need things from him. He thus projects his own desires and needs into other people, who then seem to represent to him these aspects of himself. Like the Old Man, he does not feel he is living his own life. He is always the product of his own dreams and of what he phantasizes were his father's dreams. In "The circular ruins", the Old Man is trapped in the ruins that his narcissistic endeavour brings about. In "The immortal", we can see Rufus's satisfaction at the possibility of being Homer and composing the *Odyssey* ("at least once"), of being a god, a hero, a philosopher, a demon, of being the world; it feels "divine", but also "terrible" and "incomprehensible".

Together, all these chapters highlight the profound contribution that a psychoanalytic understanding of time can bring to the understanding of the history of the individual, of historical events, and to works of literature. In this they follow Freud's path, in his understanding of the live interaction that takes place between memory and phantasy, and that finds its ultimate shape in his work on constructions, be it of an individual piece of history, or in the history of a people, as he does, for example, in *Moses and Monotheism*. Some common themes emerge, such as the notion that the realization of the passage of time is an achievement in the process of development that requires a capacity both to recognize the other, and to be able to separate from the primary objects. The contributors convincingly point out the connections between different types of psychopathology and distortions of time, and how these are reproduced in the transference to the analyst in an analysis.

As a final observation in this introduction, I would like to distinguish between the timelessness of the Unconscious, as discovered by Freud, on the one hand and another phenomenon, encountered in the patients discussed in this book, that is perhaps best expressed in terms of a "*murdering*" (or killing) of time, on the other. The former is connected to the capacity to dream, to free associate, and is at the origin of so much that is creative and on the side of Eros. The latter is the stuff of psychopathologies that are present in our consulting rooms and are a testimony to our patients' difficulties in mourning and achieving separation from their internal objects.

Note

1. This paper, originally published in *Female Experience: Three Generations of Women Psychoanalysts on Work With Women*, will appear in my volume *Space, Time and Phantasy*, to be published by the New Library of Psychoanalysis in 2008. Both books on time should be seen in conjunction with each other.

References

Freud, S., & Breuer, J. (1895d). *Studies on Hysteria. S.E., 2*. London: Hogarth.

Perelberg, R. J. (2006). *The Controversial Discussions* and après coup. *International Journal of Psychoanalysis, 87*: 1199–2220.

Raphael-Leff, J., & Perelberg, R. J. (Ed.) (1998). *Female Experience: Three Generations of Women Psychoanalysts on Work With Women*. London: Routledge.

CHAPTER ONE

The construction of heterochrony[1]

André Green

Was there ever a point in Freud's work when he was not concerned by the subject of time? One would be justified in doubting it.

Before psychoanalysis

One must begin with his works as a biologist for, in them, anatomy is considered from an evolutionary angle; more specifically, his works are situated within the perspective of the migrations of certain cellular formations in the course of phylogenesis (1990, p. 146).[2] This initial orientation was later abandoned in favour of psychopathology, but the general questions raised by his research remained present in his mind. In *Studies on Hysteria* (1895d), the idea of "strangulated affect", in other words, of the clock that has stopped, is already linked to the idea of time blocked by fixation—a movement frozen along a path evolving in time. Furthermore, in "The psychotherapy of hysteria", the chapter he contributed to the work written in collaboration with Breuer (1895d), he put forward a complex model of temporality of great originality and marvellous ingenuity. In it we can find the notions of trauma, "filed" concentric layers of memory and radial side-paths, which clearly show his concern for a complex temporal ensemble raising the hypothesis of transchronic functioning—all of which is represented from a synchronic perspective, corresponding to time in psychotherapy.

Nachträglichkeit (S.E. "deferred action")

It was in the "Project" (1895a) that Freud expounded the theory of *Nachträglichkeit* for the first time. The case history of Emma has been recounted so many times that it is scarcely necessary to return to it. Let me attempt instead to give a theoretical exposition of the situation. Let us consider a symptom formed by a constellation of characteristics (Sn). Some of these characteristics refer directly to the memory of a scene (Sc I), which sheds light on only certain aspects of the symptom. The connection between the symptom and the memory of the initial scene may be said to be preconscious[3] (conscious–pre-conscious association). Subsequently, Sc I is associated with Sc II, which occurred a few years before and was completely absent from the mind at the time of Sc I, itself occurring later in time. It can therefore be said that the connection between Sc II and the symptom (Sn) is not conscious, but unconscious. The idea that needs to be grasped here is that there is no direct link between the symptom and the unconscious memory; the latter only manifests itself retroactively by means of the preconscious memory that gives access to it. Sc I now needs to be linked up with Sc II (the first is post-pubertal and the second pre-pubertal). Sc II (enacted seduction) was accompanied by a sensation of sexual pleasure (which is why, after taking flight, the child would try to reproduce this seduction by putting herself in the same circumstances, but this time the pleasure would be sexual/libidinal). In Sc II we find some of the same elements as in Sc I, which are of secondary importance, or whose isolated and partial character do not allow elucidation of the symptom. Moreover, in Sc I there was a sexual release of the post-pubertal kind, different in nature from the pre-pubertal sexual pleasure of Sc II. The displacement of the sign of the sexual assault via the clothes on to the clothes themselves provides the explanation for the conscious association, making the latter a prominent feature of the symptom, in a rationalized form. The memory of Sc II aroused in Sc I "what it was certainly not able to at the time, which was transformed into anxiety" (Freud, 1895a, p. 354).

In his analysis of the scenes, Freud showed that only the element "clothes", common to both scenes, had entered consciousness. The associative links between the two scenes would reveal the meaningful unconscious content responsible for the late sexual discharge that continued to be linked with the memory of the assault. "But it is highly noteworthy that it [the sexual release] was not linked to the assault when this was experienced" (*ibid.*, p. 356). In conclusion: the memory induces an affect emanating from the trauma itself. Freud returned to this problem in detail in the "Wolf Man", when he discussed the date of the primal scene and its subsequent return in disguised forms. Its occurrence leaves its mark on the child who witnesses it, yet he cannot know the effect it will have on his subjectivity because he does not fully understand the way in which he has been affected by it. For this, the effects of post-pubertal sexuality at the two levels of accession to sexual maturity and intellectual development are necessary.

In 1899–1900, in *The Interpretation of Dreams* (1900a), Freud put forward the conception of the bi-directional nature of psychical processes. Progressively and regressively, cathexes traverse the psychic space between its perceptual and motor poles, which are locked, in a movement back and forth that gives rise regressively to the representability specific to dreams. The hypothesis of the timelessness of the unconscious, which is nothing more than the timelessness of its traces and of its cathexes, endowed with mobility, is already present here. This means that the psychical apparatus is caught in the double vectorization tending now towards the future, now towards the past, in the pure present of dreaming, when the flow of excitations that should lead from thought to action is impossible. Freud mentions two types of reference to time: the first recognizes the signs of its passing and draws the appropriate conclusions; the second resists them, managing not to take them into account, facilitated once again by regression induced through sleep. Furthermore, in both cases, since there is *grédience*[4] there is transition, that is to say, sequentiality capable of moving in both directions, although, as the paths of motoricity are not operative, this gives rise to *the specific mobility of* unconscious *psychic space*. In the context, the psychical processes are forced to follow a regressive path in order to be accomplished. However, this flow does not signify that meaning always follows a path from what comes first to what comes after; sometimes, says Freud, dreams even show the rabbit chasing the hunter. In any case, the situation of a chase is there. To demonstrate this, he points to the topographical regression of dreams, to be distinguished from temporal regression. It is not so much a case of returning to a constituted past as to outdated modes of expression. Hence, the twofold meaning of the denomination: they are *primary* processes, less differentiated than secondary processes, and earlier than them. If the process is called primary, it is because it corresponds to a psychic order that is considered to have existed first but which, owing to the evolution that has occurred, is scarcely present in consciousness, or so it appears. The idea that is fruitful here is that of regression, which does not occur on a massive scale, *en bloc*, but which selectively affects formal processes, arousing *fueros*, that is to say, reserves of meaning, of content, at certain moments, whereas, in other cases, it is more global, entailing an expressive and structural return resulting in a real resurgence of a time that one thought belonged to the past, as in the dynamic regression of psychosis, for example. In dreaming, although it is expressive regression (formal, concerning the form of figures and the way they are related to each other) that dominates, it cannot be said that this is the only manifestation of a temporal reference, since Freud considers that dreams refer back to an infantile scene which is modified by being transposed into a recent context. Time, however, only affects the childhood scene, referred to allusively, and its pictorial mode of expression; its actor—or its author—is not transported completely into the past. He sees the scene again profoundly disguised by the dream-work, without recognizing the sources. The reign of images is contrasted with that of language (topographical regression). The astonishing discovery was that, in their own way, images "think". Actually, dreams "neither think nor calculate" says Freud, they just transform (in accordance with wishes). But is this not to recognize that wishing is also a way of thinking—thoughts that think themselves, as Lacan said. It is the effect of a wishful world that is more at ease in the logic of thing-representations because representation can represent the wish as accomplished; that is to say, anticipation is disguised by the forms taken by the emergence of the manifest content experienced in the present. Wishful thinking cannot be separated from its accomplishment and its "realization": ". . . it [dream-work] is completely different from it [waking thought] qualitatively and for that reason not immediately comparable with it" (Freud, 1900a, p. 507).

With *The Psychopathology of Everyday Life* (1901b), Freud decompartmentalized the discovery of the unconscious: it was no longer confined to the clinical study of patients' neuroses or even to the way each one of us produces dreams under the specific conditions of sleep. Each day, the unconscious "signals its presence" in the waking life of all human beings and no longer belongs to just one register, whether of pathology or of dreams. It is present in scenes of forgetting, slips of the tongue, parapraxes, etc. There is no mention of regression here of any kind. The diachronic reference is suspended in favour of extending the scope of the synchronic register, leading to the heterogeneity of the signifier. It is a punctuation that would find its place when the problems of time were reconsidered in relation to representation. Symbolization would have the task of articulating the various registers of the signifying heterogeneity.

The return to biological foundations: sexuality

Dreaming is an experience bearing witness to the elaboration of desire, but this can scarcely be conceived without referring, in the last instance, to the activation of libidinal experience. The *Three Essays on the Theory of Sexuality* (1905d) laid the foundations for the aspect of temporality which was to enjoy the greatest success; no doubt because the iconoclastic value of Freud's text resided in its new content, but also because the thought articulating it was familiar and easily applicable. It was the thought of ordinary time apprehended intuitively, following the curve of the life-cycle which is immediately accessible to understanding: birth, childhood, puberty and adolescence, adulthood, old age, and death.

We can see how Freud's thought was swinging back and forth. Dreams had enabled him to discover a dismembered temporality, his initial intuition of non-unified time. The sexual theory returned to a time ordered traditionally, that is, in terms of the growth characteristic of life. The novelty here consisted in placing the sexual under the aegis of present time, long before its explicit manifestation after puberty, and in drawing attention to the intervention of repression in infantile amnesia, while opening the way for a return of the repressed. Thus, having returned to the long-familiar vectorized notion of time, thought was now enticed by the idea of a mode of dismantling of which it had been unaware and which was capable of challenging the idea of a past that, once past, is completely over, only reappearing under the pale hues of conscious memory. Freud was struck by the amnesia affecting the first years of life. Thus, he introduced the new category of repression, thereby relativizing his ostensible return to the notion of temporal succession assumed to govern the development of the libido. But, unlike in dreams, what has to be repressed here is neither wishes that it would have been better not to have formed, nor fantasies expressing prohibited desires, nor "wicked thoughts". Rather, it is bodily states that generate *jouissance* that is reproved, either because it disturbs the organization of the psyche or because the object involved rejects them, thereby condemning them to disappear from consciousness. These states of pleasure are due to the excitation of the so-called erogenous zones, which are highly excitable and produce pleasant sensations linked with the object of the earliest attachments, the mother, via parts of her body, that is, part-objects. In other words, there is a meeting of two eroticisms. First, her own erotogenic zones are linked up with those of the infant: the breast is placed against the mouth; then the anus is caressed by the mother while she administers bodily care, and finally the sexual organ is also aroused when the infant is being cleaned. Freud made a decisive leap here by discovering direct, immediate bodily erogeneity of an intensity that ruptures the tissue of sensible experience—a source of desires and fantasies. Furthermore, erogenous experience has its source in territories that are zones of communication between the inside of the body and the external world, where the objects which bring satisfaction are to be found (Brusset, 1992).[5]

An echo may be found here of the dual dimension of dreams, albeit organized differently: the progression of the development of the libido is accompanied by the rejection of its most intense, but also its most prohibited, manifestations, into the forgotten recesses of the unconscious. They do not remain inert there. They will be animated by an upward thrust—a sign of their vitality that has not been weakened by the passage of time. When they resurface partially—if the opportunity presents itself—they will do so in disguises that will make it impossible to recognize their origins, that is, their bodily sources bearing the marks of time. It is for the psychoanalyst's ear to retrace their invisible path. All things considered, dreams appear, then, to be *recurrent digressions*, occurring on a daily basis, since we dream every night. Each night we withdraw from diurnal temporality without being completely cut off from it, since the embryo from which dreams are formed is indeed an unconscious fantasy from the day before. The latter had already broken with the temporal flow of experience, ordered successively. It constituted itself "outside-time" (Kristeva, 1996). The space of the present instant obscures the moment in which a more or less clandestine reverie is secretly taking shape. And, similarly, once a dream has been recounted and examined piece by piece in order to subject each piece to the work of association, it will give rise to an interpretation which depends on this rupture of the time of ordinary experience. Once it has been interpreted, it dissolves into diurnal time, ceasing to solicit the dreamer's psyche once it has been deprived of its exciting, evocative power through analysis.

It is true that speaking of infantile sexuality was revolutionary in itself. Certainly, basing the idea of a subject's development on stages defined by the pleasure procured by his erogenous zones, the mouth, anus, penis, clitoris, was more than audacious; but, with respect to the temporal model, it has to be admitted that there was nothing in it to shake our conception of time, apart from the cuts of repression. Unless, that is, one turns towards the idea of layers of lava overlapping each other, causing emanations of successive temporal currents to cohabit. But here, too, the successive order is preserved. One could nuance this judgement but, essentially, time remained what it had always been. Libidinal experience conforms to a schema of development conceived of in an almost Haeckelian, cephalo–caudal manner. The migration of erogenous zones follows, as it were, the developmental line of organo–psychic growth, albeit modified by the effects of repression that blur the clarity of the temporal programme. Moreover, sexual diphasism, the effect of the combination of maturation and repression, grounds sexuality in two stages, separated by a period of latency which resists complete and full awareness—in a continuous mode—of the stages of sexuality, and breaks the continuity of memory. Latency is the period when infantile sexuality is put to rest; it settles down more than it keeps quiet, as if, following the mutative stage of the Oedipus complex, it wanted to be forgotten in order to awaken again later under the battering of puberty. The era of adult sexuality that has been inaugurated is marked, without knowing it, by the intense moments and fixations of infantile sexuality; the latter will now find that, in this new body, the consistency of its earlier organization is put to the test. Sexuality, Freud tells us, is there from the beginning. It is therefore far from being non-existent at the beginnings of life; but what he says, above all, is that it is premature. It was too early. Too early in relation to what? In relation to cultural norms, and even natural norms. The infant's body is not ready to confront sexuality, particularly with its incestuous objects. A time of waiting is necessary. When it becomes manifest, ready to act, amnesia will have covered over its first steps, which have necessarily been erased. And it would be a mistake to take these pubertal manifestations as the first ones. The essential thing is to promote the notions of fixation and regression introduced by the great precursors: Jackson (with whose work Freud was familiar, as his study on aphasia in 1891 shows), Spencer (whose philosophical influence was quite widespread among those who were concerned to construct a natural philosophy), and, lastly, in Vienna itself, Brentano.

Any kind of narrow naturalism is excluded here; for, alongside infantile sexuality, Freud shows in the *Witz*—his book on jokes dating from the same period—how the sexual drive can make its influence felt in areas quite remote from its original domain. Language does not escape unconscious causality impregnated by "tendentious" jokes, i.e., those with a purpose, which can only be evocative of the drive (as opposed to "verbal jokes"[6]). Here the pendulum swings from those manifestations that are closest to biology, on which infantile sexuality is based, to the products of language that are classically connected with the life of the mind or the soul.

Infantile sexual theories and the childhood of humanity

The analysis of little Hans' phobia would allow Freud to make a new step forward theoretically; one that the infant makes well before the psychoanalyst. I am referring to the *infantile sexual theories*. Another pole comes into play here: subjectivity, which not only has to integrate the data coming from the body (that is, biologically determined) but also the parental discourse (that is, psychically determining). Having found refuge in the unconscious, these sexual theories do not go away with age but succumb, either entirely, or in part, to amnesia. This is undoubtedly a narrative model, but one that tells the family sexual history; that of the child's sexuality as well as the imaginary history of the parents' sexuality, that is to say, the "true story" according to the child, who is aware that the essential aspects are kept from him. Both are the products of fantasy, or are even appropriated, the first on the basis of bodily experiences and the second by projection. This, then, is the first unquestionable link of intersubjectivity. It is a question of relating what parents say with what they do, which they take care to hide—only recourse to imagination can suggest solutions that may be related to infantile erogeneity. This hypothesis does not account entirely for the unconscious organization of fantasy. It needs completing by a supplementary finding: the family romance, unquestionably a historic construction rooted in folklore and fairy tales. Sexual theories have their roots in the imaginary world of the body; the family romance is rooted in the imaginary world of the hazards of destiny, which write the history that is even recounted in children's stories, explaining the reasons for a man's happiness or unhappiness. Of course, it is related history, a historical account consisting of both fable and truth. Long ways, long lies. It remains to be shown how these individual determinations echo a determining and structural pole relating to myth.

In short, Freud was concerned with the individual's prehistory, with the beliefs that the child has elaborated as answers to the mysteries of sexuality, his own and his parents', which are related to the question of his own origins. Soon after, he turned his attention to the prehistory of the species, analysing the beliefs constructed by "savages" to explain their own psyche, projected on to the external world; not only beliefs, but also acts, when they occurred in primeval times. In 1913, it was the turn of anthropology—in fact of culture as a whole—to serve Freud as a new field of exploration, enabling him to introduce what Lacan was to call the key signifiers: *Totem and Taboo* (1912–1913). The short-circuit between contemporary primitive civilizations and prehistoric societies was effected with boldness. Even though the idea of the hereditary transmission of the history (and prehistory) of civilization is not always expressed overtly, the primal myth of the primitive horde features along with its almost inevitable corollary, the murder of the father. They are related to the link between incestuous desire and parricidal wishes. According to Freud, they explain the prohibition of incest and the tribute paid to the dead father—put to death, that is, in order to appease his vengeance. But above all he offers a psychoanalytic interpretation of the origin of religions. Their influence weighs heavily even on non-believers. Irrespective of its intrinsic interest, *Totem and Taboo*, in 1913, was the prefiguration of that which, ten years further on, would give birth to the superego, undermining the foundations of the sacred, the superhuman. From the viewpoint I am taking here, it was the first stage towards overcoming a strictly ontogenetic causality that witnessed the introduction of a second pole of memory, collective rather than individual, anthropological and socio-historical rather than bio-psychological, hereditary rather than acquired. This thread in Freudian theory would never be broken, but it would only appear in a discontinuous manner. It surfaced again in 1921 with *Group Psychology and the Analysis of the Ego* (1921c); then in 1929 and 1930 *in The Future of an Illusion* (1927c) and *Civilisation and its Discontents* (1930a), finding its fullest expression in *Moses and Monotheism* in 1938 (1939a).

Marking a pause: metapsychology

The "Papers on metapsychology" (1914d), marking a time of uncertainty and integration, gathered together a good many of Freud's earlier ideas as well as opening the way for new ones. This was Freud's ultimate attempt to accomplish a theoretical integration of the first topographical model in which the various determinations were linked together on a psychical level, beyond the realm of psychology. The differences between the timelessness of the unconscious and primary processes, on the one hand, and the subjection of the conscious–preconscious and secondary processes to the effect of time, on the other, are a reminder of the contradictory aspects of temporality. They were now clarified and given a theoretical basis. But the novelty, which has not always been recognized as such, appeared in one of the final papers "Mourning and melancholia" (1917e), where he clearly introduced some original reflections on time. For the first time, Freud considered the effects of actual death, caused by the loss of the object. Whereas he had affirmed that death did not exist for the unconscious, he reconsidered the question with respect to death that is recognized by the conscious mind in mourning, while at the same time underlining the fact that the melancholic is unaware of what he has lost. He was thereby admitting that the unconscious can be affected by a loss that is experienced consciously. If the unconscious is unaware of death, it is because the desires that inhabit it are not subject to the wearing effects of time. On the other hand, the unconscious is aware of loss, not in the form of contents representing it, but because it bears its trace unwittingly, in the form of an anaemia of current cathexes. To put it in another way, the sphere of representations is not the only issue here. This was perhaps the first intuition that would lead Freud to distance himself from the unconscious and to move towards the id. For there, desires are not the only thing in question; the object, too, is in the front line. To what extent are object and time related? The loss of the object in mourning not only forces the psychical apparatus to sacrifice a part of the ego in order to make up for the void left by the loss, but refers back to an initial fixation, said to be "oral-cannibalistic", in which regression appears to touch rock bottom. The end of the object's existence—which is definitive and irremediable—takes the ego back to the very beginnings of its psychical life, in its endeavours to refind the object in its most indispensable form, at an object-less stage, as it were, when the first outlines of the relation to the object were emerging. But since this occurs in the unconscious, the ego knows nothing of this return to the past. So it brings into operation the mechanisms relevant to that time; that is, the incorporation of the object—the most ambivalent form of incorporation in which love and hate are mingled, where desire and narcissistic identification are mixed up, and where the internal and external poles can barely be distinguished.

It is clear that no form of linear development can account for the peripeteia of the theory; the Freudian pendulum continues to waver between a diachronic perspective grounded in childhood in which sexuality, unconscious desire, and now the loss of the object—a loss which used to be transitory but for which there can be no turning back of the clock now—are at stake, and a structural perspective opposing psychical systems that are organized differently. In fact the two axes, historical and structural, are complementary. That which existed at the beginning of history will constitute the primary pole, both because of its significance and because of its anteriority compared with the more developed forms known to us since the beginning of time and elaborated theoretically by traditional knowledge.

Phylogenesis and key signifiers: primal fantasies

It might be thought, then, that Freudian theory had at least now achieved a degree of stability, if not its culmination. Far from it. In a series of contributions, spread out over several years in different works, Freud was to formulate a series of hypotheses of unprecedented speculative import. The first, developed in *From the History of an Infantile Neurosis* (1918b)—the case of the Wolf Man—concerned primal fantasies forming phylogenetic patterns that function for psychoanalytic theory like philosophical categories (no doubt he had Aristotle and Kant in mind here). There was indeed a need to explain how the problems encountered in analysands, which are related to what is fundamentally peculiar to any individual temporal development, i.e., its variable and aleatory nature, could be gathered together under a set of significant characteristics making it possible to shed light on their general validity. He was to call them primitive fantasies, or rather, primal or even primordial fantasies.

What was audacious was to regard them as the organizers of sexuality: seduction, castration, the primal scene, and, last, the Oedipus complex itself. Where philosophical thought was trying to find its marks in the upper realms of thought, Freud brought the question down to the level of the basement, as he claimed with unbending prosaicness in the presence of Binswanger. From then on, Freud made constant appeals to phylogenetics and, as the latter was at odds with scientific findings, this was to irritate a number of psychoanalysts of successive generations, not to mention those who were not psychoanalysts. Some felt bound to jettison it. Some even went as far as to repudiate any sort of biologism in order to safeguard the notion of a purified psyche. The fact remains that, epistemologically, Freud had introduced a different notion of temporality, foreshadowed in *Totem and Taboo* (but there it was confined to the anthropological pole), and inherited from the earliest times of the species—a temporality passed on from generation to generation, subjecting individual vicissitudes (necessarily accidental and inconstant) to a sexual "coding" which alone is capable of marking the importance of the events in someone's life and the fantasies punctuating it. The latter, in their turn, organize the traces left by individual history. And Freud also announced a new temporal characteristic of the human psyche: *the disposition to re-acquisition*. Primal fantasies have, in effect, to be re-actualized through individual experience. I can understand that the hypothesis of phylogenetic memory traces has been criticized for being too speculative, but in that case one must answer—better than Freud did—the questions underlying it that continue to survive all the criticisms they have raised. May our modern theoreticians, who take issue with Freud's aberrant biologism, offer us better solutions that take into account the full measure of what is at stake. This is not an impossible task; but it will require an increased effort that is not simply content to erase these heuristic exigencies with one stroke of the pen.

We now know, of course, that the virus of phylogenesis had held Freud in its grip for quite some time already. The rediscovered manuscript, *A Phylogenetic Fantasy: Overview of the Transference Neuroses* (1987), which had been sent to Ferenczi, makes interesting reading from many points of view, on the condition that one realizes that it is dealing with a "fictional biology". In this respect, the person to whom it was addressed, that is, the author of *Thalassa* (Ferenczi, 1924), had little to learn from Freud, whom he was well ahead of in this area. It has to be noted that Freud never published this manuscript; he even wrote to Ferenczi saying that he could throw it away or keep it as he wished. The same was true of the "Project" which, for its part, none the less had the merit of throwing light on the sources of Freudian thought and the directions it took as it made its first steps forward. I agree that this gives us an insight into Freud's fantasies; and I also accept that certain implicit aspects of its implications are thereby elucidated. There is no justification, however, for taking a rough copy, a sort of reverie—however fascinating it might be—addressed to a more than indulgent correspondent, for a piece of well-thought-out writing, considered to be an integral part of his work. Let us not forget that Ferenczi, for his part, did not restrain any of his phylogenetic reveries. Even today, *Thalassa* still has its admirers. They seem to be delighted with the exuberant imagination of the "paladin of psychoanalysis". The censors would forgive this *enfant terrible* for his theoretical fairy tales but would be more severe with the theorizing whims of his father, Freud. There are those who will say, no doubt, that it throws a fresh perspective on the overly rational point of view that we have of the ideas of such an original author. I am far from convinced by these arguments. Most of the time I simply see them as a pretext for refusing to take other hypotheses seriously; hypotheses which are, it is true, very debatable, but whose heuristic value is none the less important.

At the level of individual prehistory, the analysis of the Wolf Man poses problems raised by the primal scene. Chapter IX of the last of the five case histories, entitled "Recapitulations and problems", shows that Freud had the feeling that he had not been very clear in his exposition of the origins and development of his patient's case. In other words, the problems connected with temporality had not delivered up their secret. The notion of *Nachträglichkeit* reappears forcefully in these reflections. We can see how the author was wrestling with the findings of the analysis and their theoretical elucidation, trying to establish the sequence of psychical events, putting forward explanations for the fixations and regressions and introducing primal phantasies as if they were philosophical *a priori*. The post-Freudian works of those who knew the patient easily persuade us that things could not be so simple.

It is worth bearing in mind that Freud carried out a painstaking historical recapitulation in a note of 1923. How can one not remain sceptical about the value of this inventory from birth up to the age of ten? And, similarly, how can one not doubt the opinion that led Freud to conclude that the patient had fully recovered when he left him? This claim is not credible when one is aware of the lamentable end to which the patient came. To put it another way, this historical reconstruction, centred on infantile neurosis, did not succeed in curing the illness, in spite of analysing the primal scene that is the pivotal theme in this text. More than ever before, psychoanalytic temporality called for other solutions. Once again, in the analysis of the Wolf Man, factual reality, ontogenetically inscribed, is opposed to fantasy—perhaps induced by the impregnation due to organizing patterns. The notion of *l'aprés-coup* (*Nachträglichkeit*) raises the question of knowing what the earlier anticipatory event, *"l'avant-coup"*, so to speak, might have been. Freud postulated the existence of a primitive knowledge—which he compared to the instinctive knowledge of animals—that would be relegated to the background in the course of individual evolution rather than being erased completely. There is indeed in man an "instinctive knowledge", referred to popularly as such, but which none the less poses a real problem concerning the nature of its function of knowledge. In certain circumstances, this instinctive knowledge may resurface and once again play an important role—its earliest origins providing the only explanation for its efficacy. Be that as it may, this earliest stage brings into play quite remarkable forms of knowing and modes of thought.

The constitution of temporality and remembering

With the *Introductory Lectures on Psycho-Analysis*, in 1916–1917, and in accordance with his plan of presenting his ideas to an uninitiated public, Freud captures, in an elegant and precise diagram, the temporality involved in the aetiology of the neuroses. He divided this temporal organization into two parts: first of all, there is the traumatic accidental event (the most recent). This only acquires meaning if one understands that its effect consists in re-awakening an earlier organization, which perhaps only exists as a sketch or in bare outline and is what he calls a "disposition due to fixation". Far from being unique or simple, the latter has two components:

(a) a sexual constitution established through the events of prehistoric life (here, Freud seems to have been playing with a double meaning: before the subject opens himself to the dimension of historicity, but also connected with the prehistory of the species);
(b) the experience of infantile life—see the *Three Essays* (1905d).

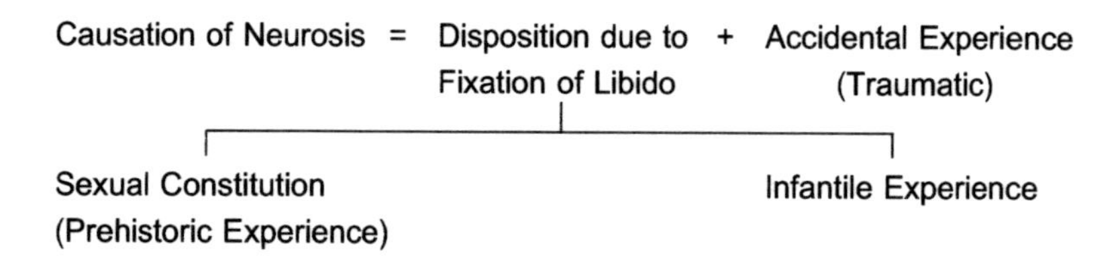

Figure 1. Taken from "The paths to symptom-formation" in: *Introductory Lectures on Psycho-Analysis* (1916–1917).

This pedagogical theoretical summary brought the existence of *diachronic heterogeneity* out into the open at last. It is constitutional, or even dependent on specific hereditary factors; in any case, it escapes the introduction of an ontogenetically determined temporality formed along the lines of programmed development. Finally, after the silence following the earlier experience —a silence that was mistakenly thought to be definitive—comes a reawakening caused by the traumatic accident that only has meaning if it is related to the earlier "disposition". Once again, what is traumatic is not the raw event, but the re-awakening of an earlier incident, which was believed to be finished with or over. It is remarkable that this diagram says nothing of primal fantasies.

Freud returns in the course of this lecture to the aetiology of trauma and the aetiology of fantasy, concluding that the difference between them is not as important as is generally supposed; what counts is the quasi-omnipotent nature of psychical reality.

However, Freud had also made another discovery, introduced in 1914 in a short article called "Remembering, repeating and working-through" (1914g), which was to steadily gain in renown. He had realized that sometimes the patient repeats *instead of* remembering. *Ergo,* repetition is a substitute for memory; that is to say, a way, which goes unnoticed and unrecognized by the analysand, of remembering a memorable psychical event, but one that cannot actually be remembered. The pendulum had now swung back to the structural pole, establishing a new relation of correspondence between memory and action. Their difference does not prevent them from being equivalent for the primary system.

The turning-point towards motion and acting out

The way was now clear for what would become known as the compulsion to repeat. *Beyond the Pleasure Principle* (1920g) theorizes this compulsion to repeat as a mode of drive functioning—of all drive functioning—owing to the eminently conservative function of the drive. It tends to be overlooked that these ideas, which Freud developed in 1920, echo in a surprising way the theories that were to emerge in France in Lacanian circles at the beginning of the 1960s. The new interest Lacan showed for structure sought its foundations in the realm of mathematics, in the work of Frege, placing the idea of "suture" in the foreground. From the moment the symbolic order insists on the relations between terms rather than the properties of the terms themselves, we are inevitably brought back to the idea of the links that are reunited by the symbolic. Synchrony is thus pre-eminent. This position brings together the thinking of Lévi-Strauss, analysing the relations between myths, and that of Lacan, highlighting the importance of signifying chains. The plunge taken by Freud in 1920 was a considerable one, since he placed the capacity for binding, which at the outset was not intrinsically linked to pleasure, before the later qualitative state marked by pleasure or unpleasure. I will return to this later in more detail.

The death drive is undoubtedly in conflict with the evolution dictated by the life drive, which also keeps the ego in a position of dependency. The temporal organization of the superego is among the most complex, since it is rooted in that of the id (itself impregnated with time that is unlinked to individual experience by means of supposed hereditary factors) and results from a split in the ego. The structure of the superego is a vehicle for influences that ensure the transmission of cultural traditions and religions. In short, it can be said that individual history is dominated at the heart of the ego by the twin determinisms of the biological heredity of the id and the cultural heredity of the superego. All this goes to show that it was impossible for Freud to accept the idea of temporality linked to individual experience alone; and, even less, the idea that the latter could be homogenous in nature, since it has to dialectalize the combined effects of nature and culture. Here the exactness of Freud's ideas is less important than identifying the main lines of his thought and the problematic issues to which they attempt to provide an answer.

The superego, the main innovation of the second topography, would like to govern the decisions of the present and, even more, those of the future. As a "protensive" agency, it serves as a guide for plans as well as actions. It is based on idealizing the past, which it sometimes renders timeless, assumed to be true "from time immemorial", raising the traditional to the level of the immemorial. The past revisited is then reconsidered and corrected from the angle of the victory that has been won over the satisfaction of the drives in the name of ethics, which, retrospectively, ennobles the aim of actions carried out under the pressure of other agencies. The superego is therefore an *orientator of time.* The directions it takes are twofold. It defines the moral imperatives concerning what is to come, so that the subject can come into being and provide, retrospectively, an ideal vision of what was, minimizing—or even suppressing from memory—the fantasized or realized transgressions in order to obtain inescapable instinctual satisfactions. Conversely, the success of sacrificing these satisfactions in favour of elevated aims is maximized. Not only the repressions but also the renunciations, and even the mortifications appeasing the sense of guilt, are glorified. In short, what tends to be erased from memory is conflict. All actions that are carried out in order to be loved by the agency symbolizing values have their roots in the spirit of the parents or in that of the lineage from which they come. Freud attributes the governing position of the agency to the *function of the ideal.*

Whatever one may think of Freud's speculations, one is bound to come to the following conclusion: the diachronic heterogeneity of the psychical apparatus is accentuated, owing to the difference of structure between the agencies and the way in which the effects of the various forms of temporality are inscribed in them. Time is now not just in pieces; its parts are in a state of tension with each other. Between the various aspects comprising it, that is to say, a time of biology marked by evolution and a time of culture marked by the history of civilizations—of their uncertain and hazardous evolution—there exists, not so much synergy, as difficulties of harmonization between its component parts, and even antagonism. One could also say that there is an opposition between the time of the subject and the time of the Other, subsuming under this term the two factors external to the ego: the deepest level inside, that is, of the psyche reaching right down into the body and the soma, and, that which is most remote on the outside, that is to say, the world and culture in which human relations play their part, working towards the discovery of otherness proper.

The Oedipus complex and castration

The heterogeneous temporal schema of the second topography leads to the crowning achievement of infantile sexuality; that is, to the stage of the Oedipus complex and then to the reorganizations that are called for by this discovery. Both positive and negative, the traces of the double Oedipus will be destroyed by the superego, leaving behind them nothing but remnants. Infantile genital organization and adult genital organization are now opposed. The differences between the Oedipus complex in boys, who overcome it by virtue of the threat of castration, and in girls, who enter it through penis envy and its symbolic equivalent, the desire for a child, again bear witness to the organizing role of sexual differences, at the appropriate time, in the integration of libidinal life.

A very important moment for theory, which appears to have no direct relation with time but in fact is closely connected with it, is fetishism, and its corollary, splitting, described in 1927. In postulating the presence of two currents of thought—one which recognizes the existence of castration and the other which refuses to substantiate it fully—Freud, who had already recognized this when analysing the Wolf Man thirteen years before, assumed that they could coexist. In fact, he defined two trains of thought which cohabit, without either of them gaining the upper hand and paralysing judgement. That is to say, without the new knowledge, founded on reason, prevailing over the past knowledge, founded on the pleasure principle. In the child's mind, two trains of thought are juxtaposed: one which not only recognizes reality as it is perceived but also the passing of time, by adhering to a conception that has drawn on the lessons of experience; and, another, which is not really repressed, and preserves intact the beliefs of the past—which still persist—concerning the mother's possession of a penis. The difference with the model of repression is clear. In the latter, old beliefs, condemned by judgement, continue to operate subterraneously (the gods from below) by co-opting each other. They emerge from their silence thanks to that which brings about the return of the repressed, as each night's dreams show. The case of fetishism presents us with a glaring contradiction, leading quite consciously to a horizontal and superficial duality, as it were. Two contradictory moments in time, one archaic and the other possessing the acquisitions of history, share between them the maternal bodily psychic space; and there is nothing to decide in favour of one or the other. My hypothesis of heterochrony is thus situated here on a strictly ontogenetic level, which can be detected in the synchrony of the present, without any reference to the mnemic-traces left by the species.

The end: the murder of the father

Now and then, Freud took the long way round. It was at such times that he spoke of the acquisition of fire and the taboo of virginity. The extravagant boldness of *Totem and Taboo* was to reappear much later on, at the end of his life, with a hypothesis that was no less surprising: the murder of Moses (already present implicitly, it is true, in *Group Psychology and the Analysis of the Ego*). Everything now came back to him at once; for instance, the hypothesis of phylogenetic traces explaining the sense of guilt that had been operative since the earliest times, as well as other ideas such as the mutation accomplished by monotheism as it became a source of progress in spiritual life.[7] Biological romanticism did not detract in any way from the importance of historical romanticism (prehistoric). Freud thought that the species also had its repressions. There is a curious point that is worth noticing here: Freud seems to forget the death drive in his explanation of the mythic and immemorial event of the murder of Moses. The death drive was already strangely absent from *Group Psychology and the Analysis of the Ego*, even though it was written one year after *Beyond the Pleasure Principle*. This is less surprising than it seems. Not all instances of violent death can be imputed to the death drive. Without saying so overtly, Freud seems to have been concerned to define *areas of causality*: the domain of the death drive is considered to be operative particularly where the biological foundations of the psyche are concerned. Where the cultural dimension is involved, other modalities come to light through *the relays* that are operative in the circumstances surrounding the events.

Freud's heritage leaves us with an uncompleted task, and we know only too well that he was constantly reworking history in all its forms. For history, he believed, could not be reduced to what is left behind it in the form of visible traces (accessible to consciousness) nor to that of which traumas conserve the memory. There is not *one* history (great or small), but *several* histories within the different spheres of the individual, culture, and the species, which are interrelated, interwoven, overlapping, and sometimes opposed—each living according to its own rhythm and its own time. All this was elaborated a long time before Michel Foucault, who was to discover it fifty years later. And rather than giving up and opting for the simplest solution—a strictly ontogenetic point of view—we should have the courage to do justice to this complexity, attempting to gather in the scattered threads of this web in order to bring together the multiple figures of time.

Notes

1. This chapter has been translated by Andrew Weller.
2. The *Lettres de jeunesse* show, through the exchanges Freud had with Brentano, how he was already interested in the subject of evolution.
3. Later, Freud would take a more radical position concerning the separation of the unconscious from consciousness.
4. Translator's note: the root in French of *progrédience* and *regrédience*.
5. Brusset, B. (1992). This book offers a remarkable treatment of all the problems raised by Freud's conception and its repercussions in post-Freudian contributions.
6. Translator's note: the allusion here is to Freud 1905c, pp. 92–102.
7. The Freudian position has been re-examined from two angles by Yosif Hayim Yerushalmi in *Freud's Moses* (1991) and very recently by Jan Assmann in *Moses the Egyptian* (1997).

References

Assmann, J. (1997). *Moses the Egyptian*. Cambridge, MA: Harvard University Press.

Brusset, B. (1992). *Le développement libidinal, Que sais-je*. Paris: PUF.

Ferenczi, S. (1924). *Thalassa. A Theory of Genitality. Psychoanalytic Quarterly*, 1938, [reprinted London: Karnac, 1989].

Freud, S. (1895a). Project for a scientific psychology. *S.E., 1*: 281–397. London: Hogarth.

Freud, S., & Breuer, J. (1895d). *Studies on Hysteria, S.E., 2*. London: Hogarth.

Freud, S. (1900a). *The Interpretation of Dreams. S.E., 4–5*. London: Hogarth.

Freud, S. (1901b). *The Psychopathology of Everyday Life. S.E.* 6. London: Hogarth.

Freud, S., (1905c). *Jokes and Their Relation to the Unconscious*. London: Hogarth.

Freud, S. (1905d). *Three Essays on the Theory of Sexuality. S.E. 7*. London: Hogarth.

Freud, S. (1912–13). *Totem and Taboo. S.E., 13*. London: Hogarth.

Freud, S. (1914d). Papers on metapsychology. *S.E.* 14: 103–108.

Freud, S. (1914g). Remembering, repeating and working-through. *S.E., 12*, p. 215.

Freud, S. (1917e). Mourning and melancholia. *S.E., 14*: 237–260. London: Hogarth.

Freud, S. (1916–1917). *Introductory Lectures on Psycho-analysis. S.E., 15–16*. London: Hogarth.

Freud, S. (1918b). *From the History of an Infantile Neurosis. S.E. 17*. London: Hogarth.

Freud, S. (1920g). *Beyond the Pleasure Principle. S.E., 18*, pp. 1–64. London: Hogarth.

Freud, S. (1921c). *Group Psychology and the Analysis of the Ego. S.E., 18*: 65–143. London: Hogarth.

Freud, S. (1927c). *The Future of an Illusion. S.E., 21*: 1–56. London: Hogarth.

Freud, S. (1930a). *Civilisation and its Discontents. S.E., 21*: 59–145. London: Hogarth.

Freud, S. (1939a). *Moses and Monotheism. S.E., 23*, pp. 1–137. London: Hogarth.

Freud, S. (1987)[1915]. *A Phylogenetic Phantasy: Overview of the Transference Neuroses*. London: Karnac.

Freud, S. (1990). *Lettres de jeunesse*. C. Heim (Trans.). Paris: Gallimard.

Kristeva, J. (1996). *Time and Sense*. New York: Columbia University Press.

Yerushalmi, Y. H. (1991). *Freud's Moses*. New Haven, CT: Yale University Press.

CHAPTER TWO

Distortions of time in the transference: some clinical and theoretical implications

James Rose

Introduction

In this chapter, I consider the distortions of time that can be observed in clinical psychoanalysis. When a patient begins an analysis that involves meeting the analyst at pre-determined points in time and for fixed periods of time, there is an impact on his external life and upon his internal world. It emphasizes something finitely limiting that may threaten various internal structures that are built upon a denial of time. Hence, from the outset, we are in a position to observe a patient's reaction to the structure of the setting that serves to organize and structure experience. The analyst's experience of the impact of the limits and boundaries on the patient come in many ways. As the transference "gathers", the patient can express this impact in verbal form; in enactments and/or in somatic experience. Indeed, my interest in time began with the everyday observation of how patients seemed to create a characteristic sense of time and space in the transference. This occurs from the beginning, but does become more focused and structured as the analysis proceeds and becomes particularly apparent when termination is a matter for consideration. But there are reasons for being interested in time other than the practical use of the denial or distortions of time as they reflect a patient's psychopathology.

In order to examine some of the implicit theoretical assumptions made about time and space in psychoanalytic theory, I have found it helpful to set some of our more familiar theory in comparison with other thinkers on time and space taken from the natural sciences.

Einstein (1920) felt that concepts of space and time emerged through the linking of direct experiences with material objects; in particular with the experience of their movement or dynamics. He felt that the concept of a material object must precede the development of notions of time and space and he was certainly *not* discussing here a theory of the development of these notions in children. His ideas enable us, as explorers of the internal universe, to wonder whether we assume particular notions of time and space when we think about the dynamics of *psychic* objects. These assumptions may also somehow bind us if we do not examine them. From here, we might ask whether looking for the distortions of time and space in the transference would help us understand a patient's psychic reality because it implies thinking about the movements of his objects in psychic space.

The Greeks distinguished two senses of time: *chronos* and *kairos*. While *chronos* refers to clock time, *kairos* refers to a sense of a special time that is significant and meaningful. It is more accurate, following the concepts due to Einstein, to think here of a space-time. We might ask whether looking for the distortions of time and space in the transference would help us understand a patient's psychic reality because it implies thinking about the movements of his objects in psychic space. Mander (1995, p. 5) recently used this distinction when discussing the significance of once weekly psychotherapy sessions in a patient's life. Thus she saw the session as a time in the week set aside from all other existence—sacred as against the profane—in which past, present, and future can come together. The obvious religious associations implied by this division of space and time imply that the separation is benign.

Nevertheless, the concept of *kairos* gives us the ability to wonder whether a patient may seek to create a space-time of *kairos* as a defence against the insistence of *chronos*. In other words, what we may be observing when we see distortions of time in the consulting room is evidence of different space-times in the patient's psychic reality.

It is quite easy to think of how the passage of time and its denial is of significance in clinical psychoanalysis. Denying the passage of time is intrinsic to resistance to change, to mourning, and ultimately death. It is also intrinsic to the denial of bodily and psychic development. Many forms of psychopathology carry with them implicit statements or theories about reality that can be thought of as "wrong". Money-Kyrle's paper on cognitive development, which developed Bion's neo-platonic theory (as Money-Kyrle thought of it) of "innate preconceptions", saw psychopathology as reflecting a patient's denial of "facts of life", among which was the passage of time. But he went further than that; he suggested that "a baby has not only to form a number of basic concepts in terms of which he can recognise these 'facts of life' but also to arrange their members in a time–space system" (Money-Kyrle, 1968, pp. 691–698).

He saw that the individual had to develop two of these main systems. One, to represent the outer world in which we have to orientate ourselves; the other develops into an unconscious system of religion and morality. Essential to the sense of orientation in either system is that it has a base, the O of co-ordinate geometry; in other words, the point of origin of the equivalent of the ordinate and the abscissa in a two-dimensional system. What is of some interest here is that the space–time system was being thought of in terms of Cartesian co-ordinate geometry. To my mind, this is a hypothesis that deserves thought and some testing. It offers a perspective on the structure of the internal world that is different from that implied by the *chronos–kairos* categories offered by the Greeks. In short, the question turns on whether a continuous time–space frame can be said to exist in the mind. What Money-Kyrle proposed was the "possible development of a kind of psychoanalytic geometry and physics with which to represent a patient's changing true and false beliefs about his relation to objects and their nature in his inner and outer worlds" (p. 694). What is implicit to this notion of innate preconceptions is a model based on what we know of human development. On the face of it, it seems to be quite acceptable that it is a "fact of life" that the prime source of all goodness for an infant is the breast, but then to assume that there is an innate preconception in the infant for this may be carrying the conclusions of empirical observation a little too far. The danger is that our conceptions become a kind of anthropomorphism so that the innate preconception becomes simply a reflection of our conception. Despite this, Money-Kyrle makes a theoretical challenge that deserves taking up.

I want to examine in this chapter the idea that the understanding of resistance to movement and development can be advanced by taking particular note of a patient's concepts of time as they are manifest in their psychic reality and, by extension, how they are expressed in the transference. I shall consider two patients where clinical issues to do with time were particularly apparent. I do not wish to imply that the only way of understanding the clinical material of the patients I will now present comes from addressing the questions arising from the patient's concepts of time. However, my understanding of the sense of these patients' psychic reality was aided by a consideration of the sense of time they created.

Clinical material

In the first case, the patient seemed to be developmentally stuck in the psychological sense in early adolescence. In other words, there seemed to be a developmental fixation. I shall seek to contrast this patient with another who showed a psychic retreat (Steiner, 1993). I think the differences between the cases lie in the sense of the first patient coming to analysis essentially feeling frustrated but with a desire for change, even if it was not clear what change was desired. The second patient sought to recreate and reinforce his retreat so that he could feel safe. At some level, becoming a patient was part of the retreat, because he used his illness as a retreat.

Patient 1. Miss Z, a neurotic young woman

> My first patient came to London to have an analysis. This move had been precipitated by a near breakdown following the ending of a love affair. The relationship had never been a sexual one. Indeed, when this patient began her analysis she was still a virgin, but well into adulthood.

A tall, rather gangling woman, she was a middle child from a small family. Her mother was described as someone who lived for her husband and found motherhood difficult and draining. Mother was also described as someone who avoided her feelings or the existence of her psychic life, although she was felt by my patient to be capable of great depth if only she would permit it. Her father was a self-made businessman who had achieved great financial success. At the outset, he was seen as an exciting man of action, intolerant of the complexities of psychic life. However, he was also seen as someone who had regarded my patient as his "favourite", which was seen as intrusive, controlling but, one sensed, secretly gratifying.

From the beginning of her analysis, time entered her thoughts because she was absolutely determined to have an analysis that would last, she thought, five years. But in regard to other aspects of her life, her capacity to procrastinate was very marked. This feature was extant on many fronts. In her professional and academic work it was constantly there, and was displayed by her inability to meet deadlines. This contained the idea of wanting to know if she was someone special to the person capable of granting a variation of the time limit. Her sense of the time limits on her capacity to bear children were to become increasingly conscious. Balanced against this was her desire for an analysis but with the knowledge that it could not be timeless. Some day, she would have to decide to return to her country of origin. The final year of the five years was conducted against the clock that in this case was not imposed by the setting but by the patient and her plans for the future, which developed in the course of the analysis.

I would like to offer some clinical material that centres mainly around a dream that was reported before a weekend. She had found that she was going to miss the last session of that week because of a planned trip abroad with a female friend. She had thought to ask for an earlier time on the Friday but somehow had felt embarrassed to raise the matter and felt that she was not explaining sufficiently what was going on. This material, therefore, provides the opportunity to see some of the anxieties and fantasies stirred by a request for a change of time, or, in other words, whether she was someone special. On the day before the Friday, she began the session with this dream.

"Last night I had this dream about you. We were in my flat and it was late at night. You thought it was about 12.30 but I knew it was 2.30 or 3.30 in the morning. It was time to go to sleep and you lay down in the alcove where my books are. The alcove was bigger in the dream than it really is and there is more room. Where you lay was separated from my bed by this trellis that my landlady has put up. Then some friends came into the room."

She did not say anything immediately in relation to the dream, but began with the difficulties that she was experiencing in talking about the problem in coming to the session the following day. She did not want to go abroad but felt that she had to, and also noted how hard it was to say to her friend that she wanted to go to analysis. I replied by saying that from the predominant feeling of the dream and from what she had just said it seemed that she regarded her analysis as a guilty or shameful secret. She said both guilty and shameful. She went on to say that she had often felt that she wanted to get up off the couch and move around and look into my eyes, but that to do so would be unwelcome to me. After some further thoughts on the theme of guilt and shame, I commented on how hard it seemed to be for her to think of coming to analysis as an act of maturity with the aim of working something out. She seemed to see it as something in which she was impossibly immature. She agreed with this, but felt that there was something about the idea of looking into my eyes that would be too daunting. I said that she was afraid of seeing rejection in my eyes. She responded by talking about a friend of hers at university who had had the ability to use her when the friend got into difficulties. The friend had been quite lacking in self-consciousness and, indeed, was unaware of the impact of her demands. She wished she could be the same.

The dominant affect displayed in this dream arose from the patient's self-consciousness. In consciousness, it seemed to arise from the difficulty in telling a female friend about her being in analysis. However, we can surmise that she was also ashamed of telling me what she wanted because of the associations concerned with looking into my eyes. Furthermore, there is a primal scene quality to the dream that came from my lying down next to her bed, with the friends eventually entering the room contributing a conscious sense of being watched. The reason that she found difficulty in telling the friend (i.e., the analyst) why she wanted to come to analysis was because the wish to come contained a sexual wish.

At this stage in the analysis, one of the most important features concerned her struggle to separate from her mother. Often the analysis was used to put her mother in the dock, and she often demanded reconstructive interpretations because they carried the promise of yet more evidence for the prosecution. Thus, it often happened that after some analysis of the transference situation between us she would then say, "What does this mean?" It usually carried the implication that whatever we had been talking about should be thought of in terms of her history. Fortunately, it became more and more obvious to her that the quest was really to reinforce the grudges and grievances that she bore towards her mother. These timeless and unending grudges, of course, bound her ever more tightly to her mother and reinforced the developmental arrest, which sought to deny the reality of her mature sexual body.

Following this period, she began a relationship with a young man of her age who was decidedly quite different from the counterpart of the sickly self-image with which she had arrived. It was a sexual relationship, but with a sense of dissatisfaction with the absence of real orgasm. She became far less prone to procrastination, and her professional life became much more successful. She had her five years of analysis and, despite all sorts of twists and turns to avoid the pain of mourning, the analysis did reach an end. It was clear that the sexual difficulties she encountered had a great deal to do with sexual feelings aroused by a paternal figure, which became increasingly conscious in the transference. This seemed to me to provide fairly conclusive evidence that the developmental arrest had held in place a phantasy of a relationship with her father before the arrival of her adolescent sexual body, which she had been very reluctant to relinquish.

Discussion

We can hypothesize that the compulsion to procrastinate was the reflection of an incestuous wish that had led Miss Z to become developmentally stuck at a pre-pubertal stage because of the fears of the consequences of having the sexual body of a woman. The fear of the acquisition of a sexual body was reflected in her history by her developing in early adolescence an eating disorder of sub-clinical severity. It was also expressed in the unrequited love affair in which she sought a narcissistic relationship that allowed her to take flight from her sexual self into a pre-pubertal body from which she could, in phantasy, remain in safety the deadly rival of her mother for her father's favours. This was also reflected in her desire to be brilliantly successful, which meant that she found it very difficult to give something up when it had been finished. It could not be given up because it was never "good enough", and therefore to give it up was to accept mediocrity. In the analysis this was reflected in her opinion that any interpretation of mine about her fear and loathing of being mediocre was in fact a contemptuous recommendation on my part that she accept that she was ordinary. This led her to be furious because I was seen as seeking to thwart her ambitions to identify with the all-powerful object that existed in her mind.

Looking now at Miss Z's experience of seeking variations in time boundaries, we can see how the picture in the transference was to seek a relationship with a powerful and impressive male figure from whom she would get undivided and special attention. Breaks and weekends generated strong reactions at the outset, as might be expected. If we look at the dream, I think we can see on the one hand a desire for the all-powerful figure betrayed in the guilty feelings about analysis. There was also a resistance to the structure imposed by the session time and a desire for it to be changed so that she could consciously feel special and at the same time unconsciously triumph over the structuring but containing maternal object of the analytic setting. Her unconscious awareness of this was reflected by the daunting prospect of looking into my eyes, or, in other words, looking at herself in the mirror of the transference. Set against this fear was the desire to be free of self-consciousness, as expressed by her unselfconscious friend who could ask for help, which meant acknowledging a wish to be in touch with her desires without the need to hide or feel ashamed of them. Thus, I propose that we can see the session and its sequence echoing some aspects of the temporal sequence of her analysis and the sequence of her life. By this I mean a developmental thrust towards an "I" in a mature sexual body, but with elements of the preceding immature phases of this sequence in close proximity. This is revealed in the dream and the associations to it. In the session the patient remains in a state of self-consciousness and seeking the approval of the analyst, perhaps envious of the perceived internal structure of the analytic setting and seeking to destroy it. However, it suggests that it is possible that in the patient's internal world there can be different time spaces that co-exist. This must mean that any psychic geometry will not be a simple Euclidean one and hence need not have single points of origin in the Cartesian sense hypothesized by Money-Kyrle, but can have many of them co-existing with each other. It also means that the implicit sense of time here is different from the *chronos* of developmental time.

It is appropriate and useful to think of this first patient in terms of fixation. The overwhelming picture was one of frustration. We might hypothesize that the compulsion to procrastinate was the reflection of a wish that had led her to become developmentally stuck at a pre-pubertal stage because of the fears of the consequences of having the sexual body of a woman. Thus, the picture in the transference was to seek a relationship with a powerful and impressive male figure from whom she would get undivided and special attention. These dynamics became acted out in professional life, but when interpreted enabled her to have the relationships more appropriate to her age. We need to remember that however unconsciously gratifying these dynamics were, they were fundamentally unsatisfying, and this realization ultimately led her into analysis. We will distinguish her from my next patient, who sought the gratifications of the retreat in a much more intractable way.

Patient 2. A patient diagnosed as obsessional

> This patient, diagnosed as obsessional, was treated in the London Clinic of Psychoanalysis. The particular form that his obsessions took was to experience episodic states of mind during which he ruminated about violent fantasies that filled him with anxiety and led him to isolate himself from other people.
>
> When we first met, he presented to me as a cheerful, personable character who was breezily good humoured. I was complimented on being an analyst "who had a sense of humour", which seemed significant at the time in that it was his parting shot at the end of the initial consultation. It seemed to carry the implication that I would be a good chap who would not disturb or upset him.
>
> An experience that he often regaled was of being rejected by an adolescent sweetheart, with whom he had had his early sexual experiences but which stopped short of intercourse. In particular, he recalled his fearful and enthralled fascination at the prospect of touching her pubic hair. She then rejected him in favour of another after about six months, when he was fourteen. He recalled seeing them together in the school yard as he looked down from a classroom, and feeling full of humiliation and a sense of bursting rage.

For the first two years, my patient attended as would be expected of a patient who was anxious to please. He appeared to be fearful of disturbing me and related in his customary cheerful but, I increasingly felt, false way. He had various ruminatory onsets and we noted that these were usually ended in response to a sense of emotional contact with me, but then this was got rid of so quickly that insight was not maintained.

In order to resist his sense of dependency on me, he acquired new ills with which I could not help but which had a symbolic relevance to his relationship with me. Very often, these took bodily form: problems with his teeth, his stomach, or his prostate. He became obsessed with the house to the rear of his own, which was reported as disturbing, enraging, and frightening him, and he became very preoccupied with his own house and anxious that it was falling to bits. Thus, the situation in the transference was transformed into bodily reality or external reality. All this was interpreted in the transference but the impact was slight, so far as could be observed.

Gradually, he lost his obsessional punctiliousness in attendance and payment, such that a constant enactment of his desire to get rid of me and treat what I had to offer with contempt was suggested. There were long periods when analysis seemed to be quite irrelevant to him, but I had a sense that he kept a very careful eye on me to see how I reacted. I do not think he was conscious of this but it resided in him as a constant state of tension and watchfulness. I also got the impression that many of his excuses for lateness and non-attendance were rather far-fetched, as if I was being tested to see if I believed them.

What this seemed to be amounting to was that the analysis was turning into one that was *endless and timeless*. However, when patients are treated in the London Clinic, there is an initial limit of three years, which can be extended in certain cases. I had obtained two extensions, but it was agreed that this was enough. Hence, we began a new analytic year in the knowledge that, so far as the Clinic was concerned, there was a known termination date. This did not necessarily mean the analysis would end, if he and I chose not to end it. I made this clear to him as we began in September. Despite this, he reacted as if he was determined not to think about it and continued on in his timeless way, apparently in state of sublime denial. This continued until the approach of the last Easter break. Then, there were signs that the prospect of a real ending was starting to become conscious. There came a day when he asked what would happen in the autumn after analysis had ended.

This session gradually ushered in a powerful onset of the ruminations. I will describe a session during this period.

The session began with the patient saying that it was as if no time had elapsed between when he had left and when he had arrived that day. His ruminations had been going full blast. Being mindful of how he seemed to attack the sense of time, I took this up with him, which met with an irritated response as if to say why wasn't I interested in his illness. Despite these protestations, I persisted with my concern about time and he replied he was always like that. He wiped things out when he left, but he added that suffering the ruminations seemed to pull people towards him. I said that his mind was full of ruminations to pull me towards him and to make sure that I would not send him away at the end of the session without having something about which to feel guilty. This led him to observe that he could always fall ill to get out of things and when he thought people were angry with him. I chose to intervene into his discursive review of his past by wondering what it was that had been going on between us that he got out of by being ill. After some skirmishing, he expressed his view that I saw him as a "little fuck" who could not do without analysis and he hated me for that. This enabled us to explore his feelings and reactions at the end of a session when I said it was time. He acknowledged that his rage vanished into ruminations; which hid his anger with me and his fear of me, which resulted from the anger in a single cut. Once in the ruminations he was ill and immune from the painful feeling arising from separation.

The anxiety that underlay the ruminations eventually began to recede as a result of his gaining some insight into a part of himself that was forever perverting ordinary affectionate feeling for men into a homosexual conception, which he began to recognize as being very deprivingly self-destructive. Initially, this was felt as a burst of anger from within his bodily self against somebody who represented this aspect of him. When he saw that this person represented a part of himself, he became capable of a deeply felt depression that seemed to be considering what could be done about how he had wasted his life. The reparative quality of this seemed in contrast to the paranoid feel of the anxieties underlying his ruminations; and had a new kind of feel to it.

Discussion

The obsessional symptoms of this patient were concerned with the experience of, reaction to, and defences against separation and ending. The experience of ending evoked such powerful anxiety that the entire conscious experience had to be obliterated and the sense of time had to go with it. This suggests that the anxiety we are dealing with here is extremely persecutory and primitive.

I have found it useful to think about the meaning of this patient's ruminatory symptoms as representing the creation of a psychic retreat (Steiner, 1993) into which he could escape from the adult anxieties of life and, in phantasy, merge with his object. The situation inside this retreat seemed to be one in which the passage of time would not be experienced in the same way as outside it. Inside, there seemed to be a timeless immobility as represented by a dream of a candle shining against a featureless background, which he reported early in the analysis. The implication is that he reacted to separation by making a cutting severance.

Fixation and retreat

With this second patient the position was more complicated than in the first, because the illness itself represented a retreat. Indeed, the very treatment could be subverted to that end. I think it may be possible to hypothesize that the fixation of my female patient for a while took the form of a retreat formed of grudges against her mother. It seems likely that this broke down in the face of recognition that she wanted children and a sexual relationship as a woman. Time, therefore, imposed an imperative upon her. There was no such imperative for my male patient, but as time has gone on he has acquired an increasing impatience with his failure to use his talents.

In their discussion of fixation, Laplanche and Pontalis (1973) say that

> fixation is repeatedly encountered as a way of accounting for a clear empirical fact, namely that the neurotic—or generally speaking any human subject—is marked by childhood experiences and retains an attachment, disguised to a greater or lesser degree, to archaic modes of satisfaction, types of object and of relationship. [p. 162]

It is a concept that Freud used at every stage of development of his theory. For example, fixation at the anal stage is said to be at the root of obsessional neurosis and of a certain character type. At the same time, fixation prepares the points to which regression will occur in all types of patient. Laplanche and Pontalis reached the conclusion that incontestable as fixation is as a *descriptive* term, it contains within itself no principle of explanation. Indeed, without other theoretical structures a degree of circularity is inevitable. Perhaps it is this attitude to the concept of fixation that can lead one to overlook some more exciting aspects of the ideas. Fixation has to be fitted at least into the topographical model to acquire any explanatory power. In their discussion of regression, Heimann and Isaacs (1952) suggest that fixation can be thought of from the progressive as well from the pathological point of view. They draw on Freud's suggestion in the Schreber case that

> The delusion formation which we take as a pathological product is in reality an attempt at recovery. . . . The symptoms of the illness are but a sign of the process of recovery which then forces itself so noisily upon our attention. [Heimann & Isaacs, 1952, p. 182]

Perhaps it is this impression of noise that marks the difference between the fixation and the psychic retreat. The psychic retreat is more noted for its silence and stability than its noise until the analysis starts to threaten it in some way. The notion of a psychic retreat can seem more complex because of the implication that it represents an organized response to something. It is a patient's creation. Indeed, we might say that whereas the fixation presents a problem for the patient, the retreat represents a solution. Steiner (1993), to whom we owe the term, saw a psychic retreat providing a patient with "an area of relative peace and protection from strain when meaningful contact with the analyst is experienced as threatening" (p. 1). I do not propose to summarize here Steiner's linking of the psychic retreat to the concept of pathological organization or the "borderline" position. What I would like to introduce here are some ideas concerning time. They derive from the observation that in both the fixation and the retreat, there is an area of the patient's personality that might be described as a space that is, to varying degrees, sealed off from the remainder. Within this space, there is a different sense of time. This sense, however, shows itself in the overall presentation of the patient such that the patient with the fixation brings a sense of urgency and demand whereas the patient in the retreat seems to bring a request for help but without the dreaded pain involved in change.

Time inside the psychic retreat

It will be recalled that in the discussion of the male patient above, it was suggested that all separation left him feeling little and powerless in a small and helpless body. It could be observed that he reacted to this by violently and sadistically severing from his object. At the same time, he projected his sadism such that he could have an idea of being a helpless victim. It is proposed that by a kind of twisting manoeuvre, he managed to be a sadist and a masochist at the same time, enabling him to have the gratifications of being the aggressor and the victim simultaneously.

It seemed to me that there was a co-existence of two processes that functioned in countervailing ways. Something was disavowed and projected into the object (in this case, aggressive impulses). Something else was gained, at the same time, from the object (in this case, the gratification of being a victim, either of the object or the illness). From the point of view of technique, interpretation of one process was stymied by the functioning of the countervailing process. It presented a formidable defence based on the determination to maintain the retreat as a closed system. It was not surprising that within a structure designed to prevent change the predominant sense of time within it was akin to that in Narnia under the rule of the White Witch — time had stopped: such that *chronos* became a sterile *kairos*. In the transference, this could be experienced as a desire for an analysis that was endless and that sought no change. The only hope of any change began to emerge when the spuriousness of these gratifications began to dawn on him.

The general conclusion we may reach from these two clinical examples is that the patients' conscious experience of, and reaction to, the time boundaries of the setting reflected very powerful wishes that could not be allowed into consciousness. These wishes not only were expressed in their reactions to the time boundaries of the setting but were strongly reflected in the pattern of these patients' lives.

Some theoretical and clinical implications

These clinical studies have certain clinical and theoretical implications that deserve discussion. Let me begin by suggesting that any theory of object relations implies the concepts of time and space. When we talk about the dynamics of physical objects, using Newtonian mechanics, we are talking about their movement in space: implying speed and acceleration, which in turn imply the notion of time. In this viewpoint, time and space provide a frame within which movement occurs. Furthermore, they are givens, and are continuous in nature. This assumption was later questioned by Einstein (1920), who showed that time and space could be neither givens nor continuous. He was discussing this in terms of the dynamics of physical objects but, it seems to me, it is also particularly true of the dynamics of psychic objects.

Freud (1933a, p. 26) made the interesting observation that, in dreams, time may be referred to in spatial terms. Thus, a dream involving small figures might be a statement about events long ago, as though looking at objects from afar. He also took the view that there was no sense of time in the Unconscious (1925a). He advanced the hypothesis that the origin of the sense of time came from the discontinuous method of functioning of the Perceptual conscious. He says, "It is as though the unconscious stretches out feelers, through the medium of the system Pcpt.-Cs towards the external world and hastily withdraws them as soon as they have sampled the excitations coming from it" (*ibid.*, p. 2). Thus, we see a sense of time coming from the iterative sampling of the external world. Here Freud is talking about time solely in its chronological or clock time sense. Within these ideas of Freud, we have two lines to follow. The first suggests that there is a link between time and space such that they are not independent of one another. The second is that there can be senses of time other than the chronological.

Marie Bonaparte (1940) summarized this work by Freud in a fascinating survey of scientific, philosophical, and psychoanalytic thinking about time up to that point. The first aspect that she discussed was the various means available to adults to mount an assault on the fact of chronological time. In so doing, she brought into focus the endless struggle between the pleasure and reality principles. In discussing the philosophical approaches, she drew attention to a remark attributed to Janet, who said,

> Generally speaking, it (i.e. time) inspires little enough affection in men's hearts, but philosophers regard it with particular loathing: they have done their very best to suppress it altogether . . . We are not surprised to find that even today [*sic*] there are patients who have a horror of time. The philosophers have felt the same. [Bonaparte, 1940, p. 454]

Why should time have engendered such horror in the philosophers? It seems to arise from the fact that time, among all concepts, can be thought of in ways that challenge the idealist and the materialist. In any event, our clinical observation forces us to recognize that there is a phenomenal experience of a sense of time and that this is separate from chronological time.

Chronological time has been linked to the notion of the "Arrow of time" derived from the asymmetry of past and future, which is akin to the Greek idea of *chronos* mentioned earlier. It is suggested by Davies (1995) that the notion of the "arrow of time" stems from the second law of thermodynamics, which rests on the observation that almost all physical processes are irreversible. Imagine, for example, trying to reverse the breaking of an egg or the mixing of milk in tea. In contrast to Newtonian mechanics, in which time becomes part of the stage of events as a result of the potential reversibility of mechanics, the irreversibility of physical processes gives rise to an asymmetry symbolized by the passage of time from past to future. From this we derive the "arrow of time" that is rendered operational by chronological time.

The second law of thermodynamics, which broadly speaking states that heat cannot flow from cold to hot bodies, can be made more precise by introducing the concept of entropy. In a simple system, such as a closed flask of water or air, if the temperature is uniform then nothing happens and the system remains in a state of thermodynamic equilibrium. The flask will contain energy, but nothing can be done with it. If heat is concentrated in a particular place, then things will happen by convection, but will reach an equilibrium and a uniform temperature. This means that the second law of thermodynamics can be reformulated into: *in a closed system, entropy never decreases*. The restriction to a closed system is important, because if heat or other forms of energy can be exchanged between the system and its environment, then entropy can be decreased. This is what happens in a refrigerator, where heat is extracted from warm bodies and delivered to the environment, but where there is a price to pay, which is the expenditure of energy.

These ideas about the open-ness and closed-ness of systems enable us to think about psychic reality comprising fields or systems that are open or closed. In closed fields, we might find equilibrium but a lack of dynamic, and the "arrow of time" does not seem to apply. Perhaps the whole point of their existence is to create the illusion of the invalidity of the "arrow of time". In more familiar words, the "arrow of time" suggests the reality of death and an interminable analysis is sought. Hence, any closed fields might be thought of as containing illusions that deaden or defend against anxieties arising from the sense of castration, the associated feeling of lack, and the connotation of death. It seems to me that this provides a link with the idea of the psychic retreat. As we have seen in the second patient described above, his psychic retreat was a closed system created to defend against the anxieties aroused by the threat of change. It follows that to seek to open up, by therapeutic action, these closed psychic systems will arouse intense anxiety that is often empirically observed. The nature of the anxiety could be seen in the second case described above, when we consider the nature of the projective processes and the implicit gratifications that created the retreat. In the broadest of summary, these entailed the gratifications of appearing to be the victim in passive submission to the aggressor, combined with the secret pleasure of being full of aggression, as revealed by the violence of the ruminations.

The clinical implications arising from this formulation are significant. Fink's paper (1993) on the treatment of an obsessional patient who showed a sense of timelessness suggested that, by careful attention to the distortion of time in the transference, significant therapeutic advances could be made. Using the theoretical terminology due to Matte-Blanco, he showed that his patient began to develop a sense of asymmetric or chronological time which enabled him thereby to have a sense of a past and a future, as well as a present. Fink noted that his patient even appeared to age physiologically. Fink suggested that since one of the functions of transference interpretations is to help a patient to distinguish psychic reality from external reality, it is this that enables the analyst to help such patients to develop a sense of chronological time. With such a sense of time comes a separate sense of themselves in the now, and in the past and the future. One has only to consider that one of the important consequences of failing to do this is to be unable to situate oneself in inter-generational history such that the passage through the Oedipal process to psychic complexity is obstructed (Bollas, 1993). Given this, we can see that consideration and interpretation of distortions of time can give significant entries into severe psychopathological processes.

Two implications of a more theoretical nature flow from these clinical studies. These concern the understanding of the psychoanalytic setting and the nature of time and space in psychic reality.

In his paper on the temporal dimension of the psychoanalytic space, Sabbadini (1989) suggested that

> The timeless quality of the content of analysis is determined by and in constant interaction with such formal time arrangements, set by the analyst and altered only by exceptional circumstances. It is this contrast of temporalities that shapes the analytic encounter, modulating its rhythm and punctuating its discourse. Each of these temporalities is unthinkable without the other. [p. 305]

Following this, I think it is possible to conceive the psychoanalytic setting as structuring and containing the patient's multitudinous fields of desire; with the patient's fields in collision with the analyst's and with the psychoanalytic setting. A visual image of such could be said to exist on many mosques in Islam; that is, the Arabesque contained by the Geodesic.

The collision of the patient's desires with the temporal boundaries of the setting permitted the observation of the first patient's desires to be treated as someone special and the second patient's desire for an interminable analysis. These essentially sexual demands are thus made apparent in the transference by the temporal structure. Analysis of the demands could then easily proceed because they are shown in an unambiguous way. We can see that in many ways, in order to be able to observe this collision of the setting with a patient's desires, we have to treat the temporal aspect of the setting as an absolute and this necessity may bring its own problems. The problems of technique often stem from the difficulty of allowing the time necessary to permit the manifestation of these processes in an unambiguous way. Premature, and therefore experienced by the patient as critical, interpretation of the patient's distortions of time (otherwise known sometimes as attacks on the setting which, of course, they undoubtedly are) may lead to a zealous compliance by the patient with the psychoanalyst's perceived demands and the essential data yielded then vanishes into this form of enactment. However, if these processes are allowed to develop, the temporal structure of the setting allows the different temporalities of the patient's internal world to become conscious and apparent to the patient and the psychoanalyst alike and therefore capable of articulation.

Several contributors to this field of investigation have suggested that a sense of time is created in the patient as a result of the session's time boundaries, which operate as a superego. It is not difficult to see that the end of a session can be experienced as an injunction: "thou shalt not . . . continue". Immediately, this injunction creates a sense of the future and, by implication, a present and a past. Kurtz's (1988) paper shows how psychoanalysts of different theoretical persuasions would treat a patient's desire to extend a session in completely different ways. By contrasting the ego-psychologist, the self-psychologist and the Lacanian views of the significance of fixed session time, he suggests that each psychoanalytical theoretical system is embedded in a metaphysics that profoundly affects the direction of treatment. Thus, the ego-psychologist attempts to strengthen a patient's capacity to distinguish and negotiate a common-sense (as he called it) reality. The framework of treatment is thought of as "real", so that the patient's distorting attitudes or efforts towards the temporal structure are necessarily "unrealistic" and therefore constitute evidence of ego failure. A self-psychologist, by contrast, might take the view that much distorting attitude reflects current ego needs and should be met by departure from common-sense reality behaviour. Thus, the analyst may allow himself and the situation to be manipulated according to the patient's needs. Kurtz saw the Lacanian system being radically different from these two in its very conception of reality. To the Lacanian, thinks Kurtz, the analyst is not present as an emblem of common-sense reality, nor does he offer himself as a self object—an illusion deemed by the self-psychologist as necessary for development. The analyst remains the impossible Other, challenging the me-connaissance of the patient's ego. Thus, we see a rationale for Lacan's short sessions being derived from these ideas and in particular for obsessional patients because of their tendency to transform the treatment into a psychic retreat. They do this by settling into a comfortable position in their analysis, where they say just enough to keep the analysis going but not enough to discover anything new. In this way the obsessional controls the analysis and can use the analyst as the ideal Other who verifies what the subject thinks and believes. In this structure, the analyst is reduced to being a mirror that simply reflects the subject's own narcissistic image of itself. There is no dynamic and hence there is no sense of time.

Turning now to the nature of time and space in psychic reality, I referred above to Einstein's thoughts on the notions of time and space. It may be thought by some that such views are irrelevant to psychoanalysis, but we should take seriously Kurtz's view that each psychoanalytic theoretical system is embedded in an implicit metaphysics. For this reason, some consideration of views from other disciplines throws our own theoretical assumptions into sharper relief.

The revolutionary consequence of Einstein's theory of relativity for the concepts of space and time was that a space-time was seen as no longer resolvable into space and time as separate coordinates. It is noteworthy that this echoes with Freud's view that space and time are interchangeable in the unconscious. In the view of Abraham (1976, p. 461), Freud made a possibly unwitting contribution to the Einsteinian revolution in the form of advocating that time and space are relative and variable. From the above discussion on open and closed systems, we can see that a model of the creation of the discontinuity of psychic space-time is the establishment of closed systems as a defence against movement and change.

One of the problems we may experience in recognizing these phenomena may arise from the difficulty in accepting the relativity of time. It is doubtless true that we gain a notion of time and space in a common-sense way as we grow up. No matter how much we seek to understand Einstein's theory of relativity, it is hard to avoid treating time and space as constants and independent of one another, as did Newton. Whereas Newton, for his purposes, was quite justified in these assumptions, if we do this then it might affect unwittingly our conception of psychic reality and, as a result, make it hard to enter and understand another's. One consequence would be to limit the view we might take of a patient's desire to "attack the setting". Our conscious minds, being rooted in the age of reason, could find it very hard to follow Freud's insight into the interchangeability of the representation of time and space in the unconscious. The fact that psychoanalysts of different theoretical persuasions take radically different positions in their interpretation of similar phenomena as seen in behaviour confirms this worrying possibility.

Being objective about internal reality involves something similar to the process of development of scientific thought as envisaged by Einstein. But, in our case, it may be more accurate to say that the observation of distortions of space and time in the transference can precede our becoming aware of the existence and nature of internal objects in the psychic reality of our patients. We can see some of the different senses of time in the psychic reality of the patients I have discussed and how this was displayed in their interactions with me. The importance of these effects was that they betrayed structures that had a profound influence on their lives.

Another important feature is that neither space nor time seemed to act as a continuous stage, but indicated that their minds could be conceived as comprising different coexisting fields in which different conditions prevailed. Empirically, this has been frequently observed upon; for example, Rosenfelt's concept of psychotic islands. These islands are, of course, islands of experience that are capable of linking to a greater or lesser extent with other experiences; sometimes not at all. The clinical studies above showed how the temporal structure of two different patients' psychic realities became apparent in the collision with the *chronos* of the setting. For this reason, the *chronos–kairos* distinction should not be seen as a dichotomous split but as two elements, struggling for dynamic equilibrium, of a dialectic in which each assumes and implies the other.

And yet, it is true that there is another, third, sense of time besides that of the chronological or asymmetric time distinct from the atemporality of the unconscious. To distinguish the two cases I have described above we can bring the notion of a developmental time implied by the maturation of the body and, ultimately, death. The impact of the reality of the limits on the reproductive life of my female patient was readily apparent in the undoing of her neurotic structures, which was not so observable in my male patient. In his case, there could be glimpses of a desire for children but this could be easily swamped by the fears evoked by the prospect of emergence from his retreat.

My reason for adding this, at the conclusion, is that it is easy to arrive at the idea that the concept of time is purely a creation of the mind; that it derives not from an actuality of experience but from an ordering of experience. Time is, therefore, to this point of view, purely a cultural creation like language.

Marie Bonaparte asked whether it was conceivable that time could be nothing but a form of her perception. She could not bring herself to believe it and brought in support of her view the psychoanalytic finding that a sense of reality and a sense of time appear simultaneously in the system of the perceptual consciousness alone. If this was the case, she argued, surely there was some link between them. What she sought to tie together here was time and reality, not in the reality principle sense, but as arising from an external reality. The aversion of men to the notion of time pointed in the same direction. Time did not arise from the pleasure principle, indeed that principle took every favourable opportunity to help us forget it. Yet, did the perception of reality arise from *a priori* in the mind, as envisioned by Kant? In a conversation after he had read her paper, Freud told Bonaparte that his views were potentially in agreement with those of Kant. The sense we have of the passing of time originated in our inner perception of the passing of our own life. With the awakening of consciousness comes the perception of this flow, which we then project on to the outside world. We are led, therefore, to a notion of a developmental time, which is rooted both in the internal and external worlds.

The observation of the existence of a developmental time complicates the picture in some ways but does offer a means of setting the struggle between the reality and pleasure principles in some context. This context seems, on the face of it, very similar to Money-Kyrle's facts of life. It is more useful to think of the *chronos–kairos* dialectic as being in dialectic itself with the arrow of time implied by the body and its maturation and the pressure that this places on psychic life, conscious or unconscious. The "facts of life" provide the geodesic within which, and against which, the arabesque swirl of conscious and unconscious life is contained and struggles. The relationship between the psychoanalytic setting and the life of an analysis is analogous, if not the same.

Summary

In summary, these observations and ideas about time and its distortion in the consulting room of clinical psycho-analysis are useful in the following ways.

1. We can learn much about a patient's psychic reality and its structure if we observe its impacts upon us. Important clues will be found in the distortions of the space and time of the psychic field between us and the form that these distortions take. Our awareness of these distortions comes in a great variety of forms and may be predominantly felt in the countertransference, very often initially in a non-verbal manner, and will take time to develop.
2. Looking at the analytic setting from the point of view of creating a time-space that contains and challenges the patient does offer a means of refining our understanding of the reasons for creating a particular setting. As an example, the reasons for treating someone five times a week as opposed to less frequently. Thus, we have a link between an understanding of the setting and the impact it will have on patients' internal psychic structures.
3. Looking at the temporal aspects of the clinical material of these two patients enabled me to grasp some important aspects of these patients' experience I could get a direct insight into their use of the analytic setting and relate this fairly directly to significant issues in their lives that were not conscious to them. We start in psychoanalysis by seeking to understand, by observation, a patient's experience and their use of the analysis. The more articulated and related these conceptual frameworks are, the more we will see and the more organized will be our perceptions.

References

Abraham, G. (1976). The sense and concept of time in psychoanalysis. *International Review of Psycho-Analysis, 3*: 461–472.

Bonaparte, M. (1940). Time and the unconscious. *International Journal of Psychoanalysis, 21*: 427–468.

Bollas, C. (1993). Why Oedipus. In: *Being a Character*. London: Routledge.

Davies, P. (1995). *The Cosmic Blueprint*. London: Penguin.

Einstein, A. (1920). *Relativity: The Special and The General Theory*. London: Methuen.

Fink, K. (1993). The bi-logical perception of time. *International Journal of Psychoanalysis, 74*: 1–10.

Freud, S. (1925a). A note on the "mystic writing pad". *S.E., 19*: 227–232. London: Hogarth.

Freud, S. (1933a). Revision of the theory of dreams (*New Introductory Lectures on Psycho-Analysis*). *S.E., 22*: 7–30. London: Hogarth.

Heimann, P., & Isaacs, S. (1952). Regression. In: J. Riviere (Ed.), *Developments in Psychoanalysis*. (pp. 169–198). London: Hogarth.

Kurtz, S. A. (1988). The psycho-analysis of time. *Journal of the American Psychoanalytic Association, 36*(4): 985–1004.

Laplanche, J., & Pontalis, J.-B. (1973). *The Language of Psychoanalysis*. London: Hogarth.

Mander, G. (1995). In praise of once weekly work: making a virtue of necessity or treatment of choice? *British Journal of Psychotherapy, 12*(1): 3–14.

Money-Kyrle, R. (1968). Cognitive development. *International Journal of Psychoanalysis, 49*: 691–698.

Sabbadini, A. (1989). Boundaries of timelessness. Some thoughts about the temporal dimension of the psycho-analytic space. *International Journal of Psychoanalysis, 70*: 305–313.

Steiner, J. (1993). *Psychic Retreats*. London: Routledge.

CHAPTER THREE

"Making time: killing time"

Paul Williams

> "Since the beginning, mankind has been submerged in a sea of time"
>
> (Hall, 1983)

We have become aware of the time-sea in which we live very slowly. The first recorded awareness of time came late in evolutionary terms, 35,000 years ago, when early modern man began burying the dead. Denotating the phases of the moon and the migration of birds, animals, and fish has always involved a temporal dimension. Recording the movements of the sun, moon, and planets and the passage of time became a basic human activity (*ibid*). Stonehenge and, later, clocks, were employed to keep a record of time. Clocks emerged in the thirteenth century as a response to the monasteries' need for accurately kept services, the nocturnal office of matins being the catalyst.

Time as an external phenomenon—chronological time—is often contrasted theoretically with psychological or subjective time: behind this is the idea that there exist two or more different types of time. What if there are no different forms of time? What if time is a unitary, unifying phenomenon and a familiar dimension of our experienced surroundings that is distinct from the processes which occur in time (Gell, 1992, p. 315)? Like its offspring, history, time is everywhere and is mediated by cultural conditions and personal psychological factors. Might the world be a big clock, albeit one which different people read very differently (*ibid.*, p. 96)?[1] Or are there different forms of time? In this chapter I discuss different experiences of time and give three examples of time from patients in psychoanalysis; each experience reflecting a different psychic state.

Observations on time

Philosophers and writers have stressed the contrasting experiences of inner and outer time and their impact. There is no space here to review these contributions, but it can be seen that there are many contrasting positions that tend to be starkly differentiated. Plato, for example, appears to have refuted the passage of time, whereas in the work of the philosopher Bergson, by contrast, there is an attempt to try to conquer the dominance of the passage of time. Similarly, Whitehead and Nietzsche held radically opposed perspectives, the former seeing time as destroying permanence whereas Nietzsche viewed permanence as an atrophying, destructive influence. Overcoming the notion of permanence and surrendering to time was Nietzsche's key to advancement personally, politically, and socially. Wherever one turns in literature—to Proust, Kafka, Mann, Huxley, Rilke, Woolf, Beckett, and many others—one encounters a preoccupation with time and the need to understand and control its influence. Most vivid perhaps is Proust's conflation of time and memory as an indissoluble unity; this union of memory with time reaches its imagistic peak in the widely used literary device of life flashing before one's eyes. Time and memory here are compressed into a single unit denoting an overwhelming experience of the impact and rapid conflation of time and subjective experience. We recognize a milder version of this experience as we grow older: time accelerates as the goals of the future become less important and the present becomes more important. The culmination of such a position leads to a perception of time as a feature of psychic reality.

Temporality is a basic aspect of experience and of consciousness in that our lives are shaped, and can only be understood, in the context of their duration, and we how grasp the parameters of ongoing experience. We have no actual sense organ with which to measure duration, and consciousness is notoriously unreliable in assessing it, especially in infancy and as we grow old, so when we speak about perceptions of time we are already speaking metaphorically (Gell, 1992, p. 93). Our grasp of time is therefore an interpretative act of the ego. To extend this view leads to a psychoanalytic perspective held by many analysts that time may be conceived of as an experience derived from a static or stable point, ultimately the self (Hartocollis, 1983). In this view, the self exists in relation to other selves or objects in a spatio-temporal framework. This is not, of course, the only psychoanalytic view, but it is a widely held one. A link to this view lies in the earliest experience of temporality, which may be the sensation of a heart beating; our mother's, our own, or both. This sequence may subsequently be detectable in clock time, in breathing, in walking, in musical composition, and even in ceremonial ritual. Behind these expressive forms of duration lies the need to feel joined up in going on being as a consequence of being connected up to objects (other people), internally and externally. The opposite experience, of being disjointed when separated from objects, is understood clinically to quickly have implications for the understanding and experience of time. The origins of this problem are not difficult to understand: for the baby, feelings of unpleasure, the inability to resolve such feelings, and their eventual relief through containment create the experience of an object that exists for the baby both inside and outside the nascent self. Separation in time or space from the object, internally or externally, initiates a primitive temporal perspective. These flows of experience come to structure the mind from the point of view of personal history, and eventually will constitute memory (Resnik, 1987). Pleasant and unpleasant experiences of duration are often projected on to the world, and this process of externalization can affect our perceptions of reality.

The experience of time as duration also implies a capacity to wait and to remember. Memory and fantasy may be seen as mediating elements in our handling the experience of duration, separation, and absence. Memory and fantasy structure the shape of psychic narratives generated to symbolize these experiences. Affects born of waiting and the ego's and superego's responses influence profoundly our experience of time. The superego's forewarning tendencies institute awareness of what may or may not happen, with all its impending consequences, whereas the need to manage normatively, repress or, if necessary, even disavow affects, gives rise to a past that can be sensed as irretrievable. Anxiety, as we know, invokes the idea that something bad may be about to happen to us; depression, by contrast, produces an experience of something bad having already happened to us. Development research into infant life has confirmed that the baby's anticipation of fulfilment involves the deployment of a hallucinated memory that protects the ego from anxieties that could lead to the break-up of the temporal sense caused by affects born of frustration and separation from the object. If the baby's capacity to hallucinate in this way fails, anxiety, fear, and anger may initiate regression to an undifferentiated "bad" self-object state in which the experience of temporality is undermined (contact with objects no longer being properly assessed) and ego activity can succumb to pre-temporal, spatial coordinates (Rey, 1994). From the beginning, the infant must deal with the temporal impact of separation and loss. A narrative constructed to symbolize the experience of being alive must, if it is to be feasible, presuppose mourning (Varvin, 1997).

Freud's time

Freud demonstrated how the unconscious has no temporal perspective, yet plays a fundamental role in the experience of time (Freud, 1915e). Its processes remain unaltered by the passage of time and wishful impulses that have been repressed appear eternal, behaving throughout life as though they had just occurred. Freud's formulation of the origin of time perception involves the system *Pcpt.-Cs* [perceptual consciousness], which is sent out to experience the external world like an antenna, returning after having sampled its offerings. This discontinuous process gives rise to a conception of time (Freud, 1925a). In his second theory of anxiety he stated that anxiety reflected adaptive efforts to deal with mental disequilibrium (Freud, 1926d). The ego, perceiving the self in danger (having cognized internal changes), mobilizes certain affects—a mental act involving consciousness. Freud was unsure about the existence of unconscious affects that exist independently and are not adaptive. The unconscious comprises *ideas,* which may or may not become affectively conscious: these are cathexes, ultimately of memory traces, whereas affects are psycho-physiological processes of discharge that express themselves as feelings. This is why anxiety can be experienced as such a powerful conscious affect, but disappears when we are no longer afraid. Freud believed that it sank into the unconscious as a latent idea. Unconscious ideas, phantasies, and feelings regarding internal objects create a template for awareness of the passage of time; a disturbed sense of time indicates a crisis in the ego's relations to objects.

Time disturbances in neurotic disorders have been widely documented. Freud's observations on obsessive patients revealed their dislike of clocks because of the certainty they imply. To remain obsessive you need to hold to an ability to doubt everything except your doubts. The obsessive's rituals and compulsive thoughts are individual time capsules, each warding off ideas of death and destruction. Fenichel noted how anal disturbances affect attitudes towards time as well as money, expressed as excessive precision or unreliability (Fenichel, 1945). Extreme punctuality and greed were linked with faeces by Jones, and are motivated, he suggests, by anal erotism (Jones, 1918). Phobias, depersonalization experiences, chronic boredom, and other symptoms reveal a disturbance in the apprehension of the experience of time rooted in anxiety about internal objects. Klein's formulation of the depressive position is, from this perspective, contingent upon a sense of time past and the impending fate of internal objects (Klein, 1940).

Experiences of time

A typical example of neurotic disturbance linked to a sense of time employed defensively is illustrated by a mildly agoraphobic patient in her thirties who is consistently late for sessions. This is followed by apologies and relief at finding me waiting. The patient complains about not having enough time. She has a fantasy that her mother and I wish to leave her. She believes I find seeing her a duty

I abhor. By keeping me waiting and finding me here she experiences the reassuring feeling that I do not intend to leave her. The idea that her mother's life, and mine, are independent of hers is disavowed, as is acknowledgement of her own rejecting feelings. I think she is also being protected from experiencing desires to be separate and perhaps sexual wishes that could signify an independent life. Censure and repression of envy, humiliation, and anger have disrupted her ego's mindfulness of its objects and her temporal judgement. Where object relations and temporality are seriously compromised, so some reversion to spatial reckoning may occur; this was evidenced by a tendency in this patient to experience, when under stress, agoraphobic anxieties.

In severe disturbances, particularly borderline conditions, the capacity to wait has not evolved as an ego function and relations with time are deeply disordered. Temporal problems take on an enveloping significance in the life of the subject, reflecting ways in which the original defect in ego integration of positive and negative experiences of objects has become an organized splitting defence against generalized anxiety (this gives rise to the "stable instability" evident in borderline thinking that, on close examination, overlays a potential for disorganization, including temporal (Kernberg, 1975)). The borderline personality could be said to live in a *quasi*-present tense, so omnipresent are anxieties concerning negativity and the fate of objects, internal as well as external. The past and the future are rendered redundant as their reality demands on the ego are felt to be unmanageable: in their place is instituted demands for immediate fulfilment as part of a chronic struggle to mitigate and avert crisis. If gratification is not forthcoming, characteristic rage may erupt, giving way in its aftermath to feelings of emptiness, despair, and, in some cases, the wish to commit suicide. Rage at not being given immediate gratification is a defence against, among other things, experiencing feelings about the past or future or both. Intense, labile affects, especially aggression, denote the loss of an object who is expected to provide instant gratification. This loss is usually internal—a psychic phenomenon—as much as it is external, and is relived via the repetition compulsion as though the original fixation had just occurred the previous moment. Alongside intense affects may be the capacity to disperse affects producing boredom, detachment, and disgust.

An extract follows from a session with a businessman in his twenties who was brought up in well-to-do but emotionally impoverished circumstances. His preoccupation with the present centred on the threat posed by the emergence of any feelings that arose in him. The following statement came at the start of a Friday session several months after starting analysis:

> P. I felt panic again coming here, coming into the room. It seems to take me over and I think it will go on forever. The thing it makes me think of is sex. It takes me two minutes to calm down. I have to make myself think of nothing, go blank and still, look at something like a plant or the door so that I know where I am. I listened to Gounod's Solemn Mass probably nine times I think between yesterday and today. I was feeling melancholic, lonely. It put me in touch with something. I felt better after a while . . . I heard from Janet [his sister] last night that she'd failed her law exams for the fifth time. I'm worried about her. How long will they let her go on re-taking her exams? . . . I had a row at work this morning. It's a complete rats' nest there these days. I'd put forward some ideas about Vancouver [a proposed business deal] but the people with the most to lose, the ones who can't stand anything new [names several names] started to try to water them down. I tried to complain but nobody took my side. It's always the same. It makes me feel hopeless and then I wonder what I'm doing there. Why don't I just leave and set up on my own? I feel I'm going on and on now and you're getting bored. I think what's worrying me most of all is Marie [his wife] going away to France for two weeks. I think she'll forget me, but I don't even know if I'll miss her.

The intensity of his anxieties, the rapid, shallow movement from one subject to the next, the transference implications regarding sexual feelings, depression at separation, the imminent weekend break after the fifth session—stand out in their significance. However, I would like to focus on this man's sense of time. He is aware, emotionally, that the weekend has arrived and he feels loss. This is disguised by the movement from topic to topic so that the separation and feelings about it cannot be properly thought. He feels panic, which he says will go on forever, yet it takes only two minutes to halt it through the manufacture of a "timeless" state. The playing of the same music induces a similar timelessness. His wife's absence for two weeks means she will forget about him entirely, which disturbs him, although he knows it isn't true. For his part he can't say whether he will notice she is gone. He is anxious about and indifferent towards his objects, a contradiction reflected in his attitude to time.

One difference between this and the previous patient is the degree of splitting employed. This is evident when my male patient shows concern about his sister and her exams (which I took to be a reference also to himself and to me). He is required to abandon the subject and speak about a row at work in which his ideas are opposed by a force that is reactionary. This force, represented by reactionary, resistant work colleagues, refers to an aspect of the patient that is averse to attempts by him to have his needs met. Such opposition to the world of objects is evidence of severe splitting and of a defensive structure that "organizes" the patient's compulsive retreat from object relationships (Rosenfeld, 1971, Steiner, 1993). After a year the patient developed further anxieties, including difficulties in being alone, driving, dealing with work colleagues, and a worry that his wife might die. To his disturbed sense of time was added problems of direction:

> P: I was anxious driving here. Marie couldn't accompany me and I felt alone, as though nobody knew me and I was invisible. It's horrible, I want to curl up and hide. I felt shaky turning right off the main road. I don't know why, it's not the driving, more the amount of oncoming traffic. It's a major junction and I don't know if I can rely on anyone letting me across, so I get anxious that I'll be stuck there. The driving is getting worse. I get confused about which gear to use, especially going up the hill to our place. I used to do it in fourth, then third but now I find myself putting the car in second, but this makes it rev and I go slower so I worry about the traffic building up behind me . . . It was Marie's birthday on Saturday and we went to a restaurant. I found the food too rich and couldn't sleep later but I'm sure it was because I spent most of my time worrying how I'd get back up the hill and what gear I should be in. It sounds ridiculous. What is strange as well as that although I want Marie with me I hate going into London with her in case we get separated. If I lose sight of her I start to feel like a speck in the crowd and I know I wouldn't be able to find my way home. I also get worried that she'll get knocked down and I won't be able to do anything.

> A: I think that when you leave here at the end of your sessions, you can feel that you have been abandoned by me, and that I no longer know

> or care about you. You feel that you cannot complain to me, even though the situation feels outrageous, and then I think you feel stuck. You struggle on, trying different strategies to cope, but I think the whole thing is feeling more and more difficult for you, and all the time you are afraid that I will find out that you feel I am responsible.

For this man, to make time for himself or someone else, in order to reflect on the past or the future, exposes him to intense, primitive feelings about those who matter to him. To accept the passage of time means being aware of and concerned for others. Making time is a depressive-position activity, necessitating acknowledgement of the other's separateness and significance to oneself and of one's role in the existence and well-being of the other person. To make time is, ultimately, an act of love. Even a solitary hobby benefits the object—in this case the self—in identification with the giver and the receiver. To be unaware of time—to kill time—is to disavow the need for the other, and the other in oneself. The borderline personality kills time to keep apart from his objects. His narcissism spares him the pains of needing, loving, and hating, even if the price may be lifelong suffering.

There are expressions of killing time that are not necessarily symptoms of borderline states but that also signify disavowal of object need. Compulsive day-dreaming, dependency on alcohol or drugs, perversions, promiscuity, or living it up in the belief that this brings happiness—all these abolish a sense of time and dependency on others. One everyday example of this is the father who, on being complimented on the beauty of his daughter, boasted, "That's nothing. Wait until you see the photographs I had taken of her" (Schiffer, 1978). Living out of time, he holds his objects at one remove, idealizing lifelessness.

Why are detective tales so popular? One reason could be that they create the illusion of suspended time: we enact passively fantasies filled with fear and curiosity, until all is revealed (*ibid.*). Similarly, Sleeping Beauty magically takes control of time, as do Cinderella and Peter Pan. And, of course, we all feel sorry for Dracula if he does not make it into his coffin by dawn to achieve his macabre victory over time.

There can be positive connotations to states of timelessness; for example, where creative contemplation may be involved. Learning to do nothing constructively (what many might call wasting time) is closely related to the concept of negative capability, to the practices of certain eastern religions, and to a mother's capacity for reverie that depends upon secure contact with internal objects. I recall a workaholic patient who reacted with horrified fascination at my comment that perhaps he longed to be able to have a session that was a waste of time—a session in which nothing substantive was achieved at all.

A kindred state to doing nothing is *reculer pour mieux sauter*—a pause or retreat into oneself to gather strength, with the help of good internal objects, in order at some point to re-enter life. These examples of apparent timelessness could be said to be appreciations of the immediacy of time and of the impact of a present tense that is lived fully.

The most dramatic reaction *against* time occurs in psychosis. The psychotic patient, as Freud observed, detaches himself from external, temporal reality and lives in an internal, timeless reality. I recall vividly a patient on an in-patient unit on which I worked who demonstrated this clearly. This socially isolated female patient was prone to suicidal ideas and had been brought up by a psychotic mother. The patient's soul had, in my view, been murdered, in Shengold's sense (Shengold, 1978). She related compliantly, convinced at a deep level that she was responsible for others' survival and well-being. While on the ward she maintained a delusion that she was healthy, married, sometimes with children, and that she was capable of predicting the future. She knew this plenitude was false but believed it to be true. Delusional controls over the past, present, and future provided a psychotic alternative to life based in a realistic appreciation of time.

Such is the primary splitting in psychosis that development can be jeopardized by the dissolution of higher psychic functions (Hughlings Jackson, 1950). The mental activity of such patients is thus often reduced to a waking equivalent of dreaming: primary process ideas and shifting identifications or part-object identifications can be taken as objects, wreaking havoc with thoughts and affects, driving the subject to feel mad. Fantasy figures may dominate the ego while actual objects are renounced and go unrecognized for what they are. In the onset of schizophrenia, particularly of the paranoid type, objects and a sense of time may be present. As time goes by the schizophrenic's mind submits increasingly to psychotic mentation and contact with objects and a sense of time diminishes, leaving the patient clinging without awareness to the present. A sense of time dissolves into more space-centred thinking, which can lead to memories being experienced as actual perceptions. Inter-object space, necessary for the use of fantasy as trial action and experimental thought, is extinguished. The psychotic person is vigilant while seemingly caring about nothing: this is nowhere more apparent than in the intense, brittle precocity of the transference. The threat of immediate engulfment coexists with unconcern. Time has vanished, yet each moment is treated as though it were the patient's last. The present is controlled in order to predict the future: killing time permits survival to the next second, the next hour. The patient's terror and vigilance engender an insane, glossolalic world of the future.

Here is an extract from a psychotic patient's analysis in which confusion regarding time is evident. Miss A had been diagnosed psychiatrically as suffering a paranoid psychosis. A fantasy object, to which she gave the name "The Director", and through which her psychosis found expression, has controlled her mental life. Miss A exploited her psychotic pathology in unusual work as an operational member of a branch of the armed forces. She feels she never had a relationship to speak of with her mother, who pushed her on to the father. She feels that she was excluded by her mother and older brother (she has two brothers, one older, one younger) and at about three of four years of age began to experience people staring and laughing at her. Her mother became pregnant with the third child and Miss A took to hiding in a shed, where she developed an elaborate fantasy life. Since this time she has been uncertain as to whether people she speaks to are real or belong to her fantasy world. She has said that she had an incestuous relationship with her father from the age of seven to fourteen. By sixteen she was masturbating schoolboys (to get love, she said) and an anorexic–bulimic cycle set in. Miss A broke down in her thirties after experiencing a fleeting wish for children. She did not behave in analysis in the way one might have expected an analytic patient to behave until many years had passed, and acting out dominated the treatment. In a Friday session from which I quote she had brought in a polaroid picture of a double bed, a matchbox to put spiders in, apparently for their safety, a pack of razor blades, a towel, tissues, and a pair of sunglasses:

> P: I just saw you on the station on the way here. Like in Westminster yesterday. You smiled at me. You looked pale so I went into a shop and bought Lem-Sip. I was hoping you'd lean on a lamp-post but when I came out you'd gone away. Do you know that people who wear hats want to get inside me? I must be a good girl, you see. I must look after the animals. I have to look after the animals. You mustn't touch barn owls, just look. Uncle Bill took me to the Albert Hall last night. It was my tenth birthday. I got a balloon. He said I'm a good girl. [Sings 'I'm a good girl, I'm a good girl'.] If I pull the string I feel it on the other end. Would you like me to do it for you?
>
> A: I think you know it's Friday today and you feel upset about the fact that I am going to be leaving you for the weekend, but I think you're being told to think of sexual things instead of thinking about what you feel. Everything's fine, let's have a party, let's have sex. But somebody is not well. I think you realize that it's you but you feel afraid of talking about it with me.
>
> P: He says cutting off my breasts and vagina makes things all right. [He being the psychotic figure who influences her thinking.] You won't hit me, will you? The dead flowers can kill me, you see. You can't do that to animals. He says you're tricking me. My tissues are broken, can I have one of yours?
>
> A: I think you're feeling so frightened to tell me about how upset and jealous you feel because one or both of us will fall apart, you fear. You feel that everything will be a disaster if you tell me the truth, that I won't be able to stand it, but the truth is you are upset, and I think that inside you want me to know about it.

The patient is out of touch with the day, year, and with my whereabouts. Her need to voice her anxieties is countered by sexualized and threatening psychotic defences. Splitting and projection on a massive scale have virtually dissolved the ego's perceptual capacity. Internal experiences are expelled with violence and are felt to take place outside her. Inside, she feels disconnected, boundaryless, timeless, and engulfed—wholly persecuted by her projections. Her awareness of time has collapsed. The struggle to symbolize without the resources to do so gives rise to unusual, poetic phrases such as "broken tissues". Spatial orientation, a developmentally earlier "container" for infantile object representations than the time-sense, is reverted to replacing the lost higher functions. Many such patients inaugurate structural conditions of relating aimed primarily at defeating psychic dissolution as much as defending against psychic pain. Within such a desperate imperative, time and separation have no psychological authorization.

The need to communicate psychic truth is what brings patients into analysis, including the psychotic. This presupposes a life with objects, in time, in the real psychological world, with the pains this inevitably brings. Appraisal of time passing is born of the ego's negotiations with its internal objects. This is implied in the developmental notion of "linear time", although this term fails to convey the many levels, digressions, and reversals that are a precondition of a concerned, depressive-position experience of time evolving. The philosopher Husserl defines linear time as "a network of evolving intentionalities", which reflects better the complex irreversibility of time. Schafer draws attention to the tragic dimension of such a view. He regards awareness of time passing as an accomplishment in the ego's system in that it demands acceptance of separateness, mortality, and the fact that every moment is different and, once past, is lost forever (Schafer, 1976). Nevertheless we try to make something of the time we have. T. S. Eliot famously echoed this in "Burnt Norton" (1944):

> Time present and time past
> Are both perhaps in time future
> And time future contained in time past
> Time past and time future
> What might have been and what has been
> Point to one end, which is always present.
> And the end and the beginning were always there
> Before the beginning and after the end,
> And all is always now . . .

In clinical work the repetition-compulsion gives emotional life a static quality based on fixation, repression and the illusional security of suffering. The analysand enacts the same ritual at the same point in time, with the same people and consequences as at the time of fixation. The existence of the repetition-compulsion implies that the past may be redone, if not undone. As Balint stated, there exists the possibility of a "new beginning" (Balint, 1952). The following simple poem (Andersen, 1995) indicates the sense of loss that may usher in temporal awareness. The sense of the meaning of time contained in the poem, although amusingly conveyed, is contingent upon mourning for both objects and for the lost opportunities development brings.

Time

We have twelve clocks in our house
still there's never enough time
You go into the kitchen
get chocolate milk for your spindly son
but when you return
he has grown too old for chocolate milk,
demands beer, girls, revolution
You must make the most of your time while you have it

Your daughter comes home from school
goes out to play hopscotch
comes in a little later
and asks if you will mind the baby
while she and her husband go to the theatre
and while they are at the theatre
the child, with some difficulty,
is promoted to the 10th grade
You must make the most of your time while you have it
You photograph your hitherto young wife
with full-blooded gypsy headscarf
an opulent fountain in the background
but the picture is hardly developed
before she announces that it is soon
her turn to collect old age pension
softly the widow awakes in her
You would like to make the most of your time
but it gets lost, all the time,
where has it gone
was it ever there at all
have you spent too much time
drawing time out

You must make the most of time, in time,
roam around for a time without time and place
and when it's time
call home and hear
"You have called 95 94 93 92?
That number is no longer available."
Click.

Time is omnipresent and a central human preoccupation. Varying definitions of time arise as a consequence of cultural and psychological influences reflecting subjective and social perspectives. Nowhere is variation in awareness of the sense of time more apparent than in psychoanalytic treatment where the conditions of subjectivity are manifest. Differences in the experience of temporality are related to intrapsychic relations between the ego and its internal objects, in particular the capacity of the ego to tolerate and mourn separation and loss as durational phenomena—an essential aspect of development. The three clinical examples given illustrate increasing disorientation regarding time in parallel with levels of disruption to internal object relationships. Defensive, processual, and structural levels of disturbance reveal a progressive inability to mourn, and this is reflected in disturbances of time perception. The examples given from poetry reflect this psychoanalytic premise.

Note

1. Americo-Indians, African tribesmen, and other third world groups can treat time with exceptional patience compared, for example, to westerners. All concur that time exists, but its salience in our affairs is different.

References

Andersen, B. (1995). *Cosmopolitan in Denmark, and Other Poems About the Danes*. Copenhagen: Borgens.

Balint, M. (1952). On love and hate. In: M. Balint (1965) *Primary Love and Psychoanalytic Technique* (pp. 141–156). London: Tavistock.

Eliot, T. S. (1944). Burnt Norton, in *Four Quartets*. London: Faber & Faber.

Fenichel, O. (1945). *The Psychoanalytic Theory of Neurosis*. New York: Norton.

Freud, S. (1915e). The unconscious. *Standard Edition, 14*: 159–215. London: Hogarth.

Freud, S. (1925a). A note upon the "mystic writing pad". *S.E., 19*: 227–232. London: Hogarth.

Freud, S. (1926d). Inhibitions, symptoms and anxiety. *S.E., 20*: 77–172. London: Hogarth.

Gell, A. (1992). *The Anthropology of Time*. Oxford: Berg.

Hall, E. T. (1983). *The Dance of Life: The Other Dimension of Time*. New York: Anchor Press/Doubleday.

Hartocollis, P. (1983). *Time and Timelessness*. New York: International Universities Press.

Hughlings Jackson, J. (1950). Factors of the insanities. In: *The Selected Writings of John Hughlings Jackson*. London: Staples Press.

Jones, E. (1918). Anal–erotic character traits. In: *Papers on Psycho-Analysis* (5th edn). London: Bailliere, Tindal & Cox, 1950.

Kernberg, O. (1975). *Borderline Conditions and Pathological Narcissism*. Northvale, NJ: Jason Aronson.

Klein, M. (1940). Mourning and its relation to manic-depressive states. *International Journal of Psychoanalysis, 21*: 125–153.

Resnik, S. (1987). *The Theatre of the Dream*. New Library of Psychoanalysis, No. 6. London: Tavistock.

Rey, H. (1994). *Universals of Psychoanalysis in the Treatment of Psychotic and Borderline States*. London: Free Association.

Rosenfeld, H. (1971). A clinical approach to the psychoanalytic theory of the life and death instincts: an investigation into the aggressive aspects of narcissism. *International Journal of Psychoanalysis, 52*: 169–178.

Schafer, R. (1976). *A New Language for Psychoanalysis*. New Haven, CT: Yale University Press.

Schiffer, I. (1978). *The Trauma of Time*. New York: International Universities Press.

Shengold, L. (1978). Assault on a child 's individuality: a kind of soul murder. *Psychoanalytic Quarterly, 47*(3).

Steiner, J. (1993). *Psychic Retreats*. New Library of Psycho-analysis No. 19. London: Tavistock.

Varvin, S. (1997). Time, space and causality. *The Scandinavian Psychoanalytic Review*, *20*: 89–96.

CHAPTER FOUR

Existence in time: development or catastrophe?

David Bell

There is something very peculiar about the representation of time. It appears to us as both a creation of our minds and yet independent of us. Once we have it, we cannot *not* have it. We cannot conceive of a world in which time does not exist, and yet the measurement of it is arbitrary and man-made. Because of its peculiar status, felt as not quite internal yet not quite external, it is, like death, easy for us to project on to our representation of time certain persecuting ideas—we talk of "killing time", being "trapped" in it, "time catching up with us". It is a rare occurrence for us to locate in the sense of time feelings of joy or peace, these being more commonly associated with states of timelessness.

Awareness of the passage of time is inextricably linked to thoughts and feelings of mortality, the transience of all things, which Freud addresses in his short paper "On transience" (Freud, 1916a), which is central to the theme of this chapter and will be discussed in more detail.

The title of this chapter, "Existence in time: development or catastrophe", serves to express the idea that feeling oneself as existing in time is an important developmental achievement. For some, however, it is felt as a permanent imminent catastrophe evaded by the creation of a timeless world where, apparently, nothing ever changes, an illusion of time standing still. However, the attraction of the illusion is undermined by the ever-present terror of expulsion from this world, a "Garden of Eden", precipitating a situation that brings the possibility *not* of development but instead a sense of sudden deterioration and death. It is as if all the time that they have managed to ignore suddenly catches up with them. They feel they will suddenly grow old without ever growing up.

I will begin, however, by discussing not a patient but a character in a novel, who sells his soul to remain forever young and suffers the catastrophic consequences of this Faustian bargain. His very method of evading the (to him) catastrophe of growing old is ultimately the source of the actualization of the very situation that fills him with horror—namely, sudden deterioration, ageing, and death. I am referring to Dorian Gray, the eponymous hero of the novel by Oscar Wilde.

Synopsis of The Picture of Dorian Gray

The novel opens by introducing us to the central three characters: Basil Hallward, an artist driven by high romantic ideals, his friend, Lord Henry Wotton (Harry), a highly intelligent cynic much admired by his circle for his cleverness (Wilde emphasizes his capacity to turn everything on its head), and our eponymous hero, Dorian Gray, a young man of "exquisite beauty", a naïve romantic and innocent in the world. Basil Hallward has been inspired by Dorian's beauty to paint a picture of him, which he regards, and Harry agrees with him, as his finest work, the realization of something towards which he has been striving all his artistic career. Despite his belief in the level of his achievement, he resolves not to exhibit his picture as, "I have put too much of myself into it" (p. 9). Interestingly, and I will return to this later, Harry describes the figure in the portrait as "Narcissus".[1]

Dorian, befriended by Basil, is soon introduced to Harry. Harry has a strangely seductive way with words. He is invited to all the best dinner parties; people hunger to be liked by him. Harry is a Mephistophelean figure and his clever philosophizing serves as a lure into a world where nothing matters, there are no values—all is cleverness and repartee: "I choose my friends for their good looks, my acquaintances for their good characters, and my enemies for their good intellects" (p. 15). "The one charm of marriage is that it makes a life of deception absolutely necessary for both parties" (p. 10). He believes in no genuine goodness or depth in others, claiming it all as hypocrisy: "We praise the banker that we may overdraw our account, we find good qualities in the highwayman in the hope that he may spare our pockets" (p. 88). "We think that we are generous because we credit our neighbour with possession of those virtues that are likely to be a benefit to us." He indeed gives substance to aphorism: "The devil has all the best tunes".[2]

Basil feels his life has changed since meeting Dorian, the realization of his artistic ideal, "my life as an artist depends on him". He fears Harry will exert a bad influence on him—this fear is well-grounded; Harry soon has Dorian under his spell. On their walks, Dorian is soon captivated by him and hungers for his "philosophizing". Harry talks of the need to take from life all it can offer, surface is everything, human depth a mere comforting fiction: "it is only shallow people who do not judge by appearance" (p. 30), and through his talk he instills in Dorian an awareness of his (that is Dorian's) beauty and a terror of the passage of time, ageing, and death. As he stands in front of the finished picture, it is not Basil's artistic depths that stir him but Harry's "panegyric on youth, his terrible warning of its brevity" and, as the full realization dawns on him, the passage continues:

> Yes, there would be a day when his face would be wrinkled and wizen, his eyes dim and colourless . . . the life that would make his soul would mar his body. . . . As he thought it, a sharp pang of pain struck through him like a knife . . . He felt as if a hand of ice had been lain across his heart. [p. 33]

At the same time as he is suddenly struck by is own mortality, he has the realization that the picture, in contrast, is immortal and so will, so to speak, defy time.

And then he utters the prophetic words that pre-figure the Faustian deal:

> If it were only the other way! If it were I who was to be always young and the picture that was to grow old! . . . Yes, there is

> nothing in the whole world I would not give! I would give my soul for that. [p. 34]

Dorian cannot conceive of the possibility of growth bringing development but only loss, deterioration and death. Without realizing it, he has dammed himself, sold his soul to the devil, and with it his whole sense of reality.

Basil warns Harry not to say things before Dorian, not to contaminate him with his poisonous ideas. But it is too late. Harry replies, "Before which Dorian? The one who is pouring the tea for us, or the one in the picture?" (p. 37). Dorian is already altered.

What we subsequently witness is a struggle in his character between a better part of himself, which retains a sense of what is real and good, and another part of himself driven towards corruption and perversity—a world where human values and the realities of existence cease to have any dominion over human affairs.

At this point, however, Dorian still retains much of his innocence and also his belief in beauty and the power of love. He falls in love in the most passionate and romantic way with a talented actress whose performances he watches, enthralled, every night. Eventually, he goes backstage and declares his love for her. The actress, Sybil, leads a terribly unhappy, impoverished life; she has nothing except her art, which is all to her, and she pours her whole life into it. She falls passionately in love with her "Prince Charming", and they are engaged to be married. Dorian and Sybil are joyous, expectant lovers and, from the structure of the story, it is clear that love might have saved Dorian from his fate. But it is not to be. Dorian takes Basil and Harry to watch his betrothed in a performance of Romeo and Juliet. Sybil's performance, however, is a disaster and Dorian is humiliated in front of his friends. When he goes to see her backstage, she acknowledges that, having now found love in reality, she has lost her interest in the third-rate theatre, with its retinue of poor actors, that had been her life: "You made me understand what love really is." So, the very point where Dorian is rapidly losing touch with reality, sacrificing it for a wish-fulfilling illusion, Sybil is giving up her deep-felt joy in the depths of artistic creation viewed now as shallow in contrast to the greater "reality" offered by her lover.[3] "My love! My love! Prince Charming! Prince of life! I have grown sick of shadows. You are more to me than art can ever be" (p. 101).

But Dorian no longer wants her if she cannot produce a grand performance to impress his friends, and spurns her love in a most cruel way (reminiscent, perhaps, of Narcissus spurning Echo) and, as she begs him not to leave, his voice, the reader realizes, is no longer his own but that of his mentor, Harry, mocking her passion, her depth of feeling. The writer echoes his cynicism, "There is always something ridiculous about the emotions of people whom one has ceased to love. Sybil Vane seemed to him to be absurdly melodramatic. Her tears and sobs annoyed him" (p. 103). In his cruel mockery he feels no guilt.

This is the central moment of transformation. As he wanders the streets all night, one senses that his character is now irredeemably altered. When Dorian returns home, the writer stresses the *material* wealth of his surroundings (contrasting with the spiritual poverty), "his new-born feeling of luxury" in the home he has just decorated, "the huge gilt Venetian lantern . . . the great oak-paneled hall, the renaissance tapestries" (p. 105), but as he surveys all its splendour, his eye falls on his portrait and finds that the face had altered, bearing the mark of his changed character: "The quivering, ardent sunlight showed him the lines of cruelty round the mouth as clearly as if he had been looking into a mirror after he had done some dreadful thing". When he looked at himself in an actual mirror, there was "No line . . . that warped his red lips" (p. 105).

As Green (1979) comments "the portrait clearly will be the visible index of his morality" (p. 394).

Awaking from sleep, Dorian recaptures some of his sense of reality and, feeling deep remorse, endeavours to make amends. When Harry arrives he anxiously tells him of his plan to try to repair the damage he has caused Sybil. Harry, however, informs him that it is too late; the morning's newspapers all carry the story of Sybil's suicide—his terrible acts are already beyond reparation. Harry, showing the depths of his cruelty and cynicism, congratulates him, telling him what a fine thing it is for a young man to have a woman kill herself out of love for him. "I wish", he says, "that I had ever had such an experience" (p. 118). Dorian is seduced and quickly released from all guilt.

> "So, I have murdered Sybil Vane . . . murdered her as surely as if I'd cut her little throat with a knife. Yet the roses are no less lovely

for all that. The birds sing just as happily in the garden. And [addressing Harry] I am tonight to dine with you, and then go on to the opera, and sup somewhere . . ."

And in a perverse reversal of depth and artifice he exclaims:

"How extraordinarily dramatic life is . . . if I had read all this in a book, Harry, I think I would have wept over it. Somehow, now that it has happened actually, and to me, it seems far too wonderful for tears." [p. 115]

Yet Dorian is not completely taken over by the perverse distortions and shows some insight when he states:

"Harry. . . why is it that I cannot feel this tragedy as much as I want to? I don't think I am heartless. . . . I know I am not. And yet I must admit that this feeling that has happened does not affect me as it should. It seems to me to be simply like a wonderful ending to a wonderful play. It has all the terrible beauty of a Greek tragedy, a tragedy in which I took a great part, but by which I have not been wounded." [p. 117][4]

Dorian's life disintegrates. He lives a life of boundless and guiltless pleasure. He frequents brothels, consumes drugs, has numerous relationships, and is responsible for ruining many peoples' lives and for the deaths of others. But his *appearance* is untouched, remaining forever youthful while the portrait bears all the signs of his cruel and debauched life. The portrait bears witness to his crimes, carries the signs of ageing and the deterioration of his character; he cannot bear to look at it and hides it away in an attic.

Basil (the artist) comes to see him, dismayed by what he has heard of Dorian's life. His is the voice of morality and conscience. Dorian is, however, unmoved, and takes him up to the attic to show him the portrait now so transformed that even the artist can hardly, at first, recognize it. But even now, the writer makes clear, all is still not lost. Basil can still see "the horror of whatever it was had not *entirely* spoiled that marvelous beauty" (p. 179, my italics).

Basil, realizing that the picture confirms the evil that has dominated Dorian's life, suggests they pray, but Dorian mocks him and then coldly murders him. He later employs a scientist, whom he blackmails into using his chemical skills to get rid of all traces of the corpse. The last possibility of redemption now lost; Dorian's life deteriorates further.

Towards the end of the book, Dorian comes near to facing all the evil of his life and, through concern, he spares a woman from becoming entangled with him. He then mocks himself and sees this only as hypocrisy (as if echoing again Harry's lack of belief in anything good). Finally, at the point of recognizing the horror of his life, he has no belief left in any goodness within himself. He climbs the stairs to the attic to confront the portrait, "his own soul . . . looking out at him from the canvas and calling him to judgment" (p. 139). In desperation, he tries to rid himself of this awareness of reality, felt as an unbearable persecution, the sight of his own self aged and ugly beyond belief. He grabs a knife and lunges at the portrait in an attempt to destroy it forever.

The servants hear the noise and when they go up to the attic, they pass the portrait, now transformed back to its original youthful beauty. On the floor is a man with a knife in his heart. The body "was withered, wrinkled and loathsome of visage. It was not until they had examined the rings that they recognized who it was".

Discussion of narrative

There is something about this story that grips us; it has the same qualities as the great myths and fairy stories that speak so directly to universal human themes. The core of the narrative is the deep understanding the author brings to the cost to character of evasion of the facts of life, which include the inevitability of ageing and death, the feelings of guilt that are part of life and that lend it its moral force—all of which are supported by the awareness of the passage of time. It is clear that for Basil, the artist, who represents throughout the novel a moral force, the passage of time is an inevitable aspect of human experience that he can accept, supported in this through his capacity to symbolically capture lasting beauty in his work, while accepting its passing in life, and this gives his character depth. Harry, however, occupies a completely different position. For him, there is no real meaning to life, any claim to such is mere hypocritical posturing; all is surface, there is no depth. His philosophy is supremely narcissistic—the aim of life is to have as much pleasure as possible, regardless of the consequences for others. He is cynical of any views to the contrary (e.g., "We praise the banker in order that we may overdraw our account, we find good qualities in the highwayman in order that he may spare our pockets"). Dorian appears first as an innocent in two senses: he has not yet understood the inevitability of the realities of life, particularly of time passing and the process of ageing, and, further, he is innocent in the sense of not having been corrupted by experience. Emergence from innocence, the gaining of knowledge of life, *might* have brought ordinary sadness and therefore depth to his character, but Harry succeeds in creating in him a perception of time that brings only of horror; there is nothing to be gained from growing older—only loss of the only thing that counts, surface beauty.

Dorian's solution to this painful conflict is the perverse solution. Rather than adapting to reality, reality itself is altered; rather than bear the pain of the loss of his ideal self, he preserves it forever by exchanging places with the portrait—he will remain forever young, while the portrait will bear all the marks of time's passing. His exchanging places with the portrait, an act carried out perhaps initially with the aim of preserving life, because of the stasis that it brings to development, is in reality a psychic death, represented in the story by the gradual deterioration of Dorian's character, his inner reality, the cost he pays for preserving only the surface appearance of his existence. (See Harry's cynical defence: "People say sometimes that Beauty is only superficial. That may be so. But at least it is not so superficial as Thought is. It is only shallow people who do not judge by appearances" (pp. 29–30.)

What, then, is offered in the narrative is denial as a solution to the problem of loss, a denial, however, that, as so often in life, is never fully successful. For Dorian is constantly persecuted by the awareness of the passage of time as represented by the changes in the portrait, hidden away in the attic but never—psychologically—out of view. Although apparently enjoying all the perverse experiences that come his way, Dorian still has always to return to his home and his awareness of what is hidden away in his mind, symbolically in the attic, that can never be fully faced nor fully erased. At the end of the story, he tries finally to destroy the portrait that bears witness to the unbearable reality of his life; but it is his own self that is destroyed. The author here, I think, shows a deep insight into the underlying motives of some suicides, namely the delusional idea that one can be rid of some unbearable inner object and live on free of it.[5] Dorian suddenly grows old and dies, and this can be taken as representing symbolically the collapse of an illusion of timelessness experienced in this terrible way, as if all the time he has denied suddenly catches up with him in one catastrophic moment.

Transience: Freud, 1915

In 1915, at a time when Freud was much preoccupied with the question of mourning and with the human horror of the Great War, he wrote a short paper, "On Transience", in which he describes a walk in the countryside "in the company of a taciturn friend and a young but already famous poet". His companions could not properly enjoy the beauty of the scene as appreciation of it was coloured by painful thoughts arising from awareness of the passing of all beauty and the inevitability of decay. Indeed, had they been contemplating Dorian's portrait, and not the beautiful mountain scene, they might have echoed Dorian's words "I am jealous of everything whose beauty does not die. I am jealous of the portrait you have painted of me. Why should it keep what I must lose" (p. 35).

Freud could not share the gloom of his companions and, with great conviction, points out that the ephemeral nature of the beauty does not detract from its value but, on the contrary, adds to it: "Transience value is scarcity value in time . . . we would [not think] a flower that flowered only for a single night . . . on that account less lovely" (Freud, 1916a, p. 306). The fact that all things must pass and that even all animate matter must cease to exist should not detract from its emotional significance of beauty in the world "since the value of all this beauty and perfection is determined only by its significance for our own emotional lives, it has no need to survive us and is therefore independent of its duration" (*ibid.*).

Freud's companions were unconvinced by his argument and remained with their gloomy thoughts, which led Freud to suggest to himself that they must have some emotional difficulty that stands in the way of this simple understanding. This "emotional difficulty" bears some important similarities to that faced by

Dorian: they can only conceive of time passing as persecution. Awareness of the transience of all things catches then the touchiest point in our narcissism—the awareness of mortality.

'On transience' was written earlier in the same year as "Mourning and melancholia" (Freud, 1917e) , where Freud discusses the difficulty the ego has in giving up its lost objects and breaking its attachment to them, a long and painful process. When referring to loss, he was not referring only to the loss of people, but also of the loss of more abstract qualities, such as one's ideals. However, from the merely economic standpoint, it was hard for Freud to offer any reason as to *why* the work of mourning is always so arduous—apart, that is, from making the rather general statement that the ego is very conservative in accepting the loss of anything that it values, and seeks to preserve it in one way or another. From this perspective, the struggle of mourning is in essence a struggle between the pleasure principle (which denies the reality of loss) and the reality principle. Ultimately, in the right circumstances, the reality principle holds sway.

All this arduous work might appear to be wasted energy—how much more practical, one might think it would be if the human mind, when faced with the reality of the loss of its loved objects, could immediately give up all attachment to them, de-cathect them, and replace the loss with an object that is available. Yet, if in life we met such a person, someone who immediately lived his life such an "economic way", who, when the object of his love ceases to be available to him, gives it up with great facility transferring his attentions elsewhere, we would regard such a person as shallow, lacking in character (in fact, just like Harry in *The Picture of Dorian Gray*). For we recognize that the pain of mourning is not without purpose —it brings depth to character. The *apparent* freedom to replace attachments with such facility would in fact be an enslavement to narcissism. Freud, however, has no place for such an understanding in his metapsychology. But, as is so common with Freud, when he recognizes something to be true but lacks a theoretical framework in which to house it, he turns to a more literary form. In the brief paper "On transience", Freud introduces a dimension to the argument that he could only express in this more literary form; for here he argues that the capacity to really enjoy nature without being persecuted by the awareness of the ephemeral nature of all life is underwritten by the *ability* to mourn loss, including the loss of the conception of oneself as immortal. *The capacity to mourn* is here viewed as an achievement of the ego; it brings aesthetic depth and pleasure. It is of interest that one of Freud's companions, who had such difficulty with mourning, appears to have been one of the greatest poets of his century, Rilke (Gekle, 1986).[6]

Freud, in making this crucial link between the capacity to mourn and the capacity to accept one's own mortality, anticipates Melanie Klein's description of the depressive position, where the capacity for mourning is linked with the capacity to bear the pain of awareness of separation and guilt. Before discussing this further, I will discuss some material from the analysis of a patient, who is in some way typical of patients who are forced to evade the pain of awareness of the passage of time through maintaining, often in a hidden way, an illusion of timelessness. They are always about to make some crucial developmental step but remain incapacitated from doing so. There is usually a history of profound emotional deprivation in childhood, which leaves them prey to catastrophic anxieties, against which they have to build a powerful defensive organization.

Clinical material

Mrs S, a woman of forty-one, came to analysis in a way that is typical of the patients I am describing. The illusory world that she had created had been unable to withstand the effect upon her of some catastrophic losses in her life. She moved rapidly from a position in which, as I subsequently learnt, she felt as if she had managed to evade the ordinary blows that life inevitably brings, such as awareness of dependence on others (felt to be a reprehensible and terrifying state), the consequences of loss and separation. She told me that she cried ceaselessly for two years and then "suddenly grew up". In other words, in the place of emotional development in which her place in the world could be faced and worked through, she created instead an alternative world which evaded the features of reality that she could not manage. She was much supported in this by her family configuration. She has two brothers one or two years older, and one sister four years younger. However, as she understood it, she displaced all her siblings and her father in her mother's affections, to the extent that, as she viewed it, the primary couple was herself and her mother, father and siblings appearing as envious children bearing all the consequences of being excluded. She was excessively preoccupied with her appearance, wishing to remain forever young. However, the traumatic losses had precipitated her out of this world into a catastrophic situation; she felt her body was disintegrating, that she might die at any moment. Though manifestly seeking urgent help for this situation, she inevitably used the analysis to restore her equilibrium, recreating in the analysis the illusory world that preceded the breakdown and organizing her life in such a way as to support her in this state. Before starting analysis, she married precipitately. Her husband brought her and collected her from every session. Though at first this was necessary, given the very disturbed state she was in, she seemed to lose sight of this, and it became in her mind something quite different—not a vulnerable woman being collected from analysis, because she found it so disturbing, but more as confirmation of a privileged position in life; she was always picked up, so that she never felt dropped. In the same way as she felt that she could evade the ordinary (though for her extraordinary and persecuting) consequences of being human, she felt she could avoid the consequences of being an ordinary patient—bearing the various pains and frustrations that arise, inevitably, from being in analysis. I, as her analyst, occupied the same position as her husband, providing a sort of home or "psychic retreat" (Steiner, 1993) for her to live in, protecting her from life and making no demands. Although *apparently* happy in her illusory world, it was very clear that she felt an almost constant persecuting anxiety—she was hypochondriacal, and dreamt frequently of being attacked. She ceased being able to have a sexual relationship with her husband. She was, however, paralysed in a timeless world and the threat of any action that would have real consequences precipitated her into a terrifying world that she could not control. Deepening of her relationship with her husband brought to mind the possibility of intercourse and pregnancy, the latter dreaded as the paradigm of something growing inside that could not be stopped. This kind of anxiety became manifest in the analysis in her determination to control it, while constantly being threatened with the idea that she could not.

A related difficulty was her paralysis in making any decisions, as to make one decision was, necessarily, to exclude the alternative and so bear the pain of that loss. Her life was therefore dominated by a kind of inconsequentiality.

This was realized in the first period of the analysis by its endless repetitious quality, every session starting in the same way with her expressing her belief that I was trying to humiliate her and force her into a position of dependency to gratify my perverse needs. Often, however, by the end of the session, a different picture of me emerged, a figure with whom she felt safe. It took me quite a long time to realize the extent of this problem, and when she and I were able to discuss it more directly, she told me that she *never* thought about her analysis between sessions. She desperately wanted to be cured of her constant state of persecution, but wanted it to be painless; that is, she wanted the results of analysis without having to be really involved in it. It was, of course, always very difficult to distinguish between a genuine dread of being brought into contact with reality—felt as threatening her with collapse—and a demand that I endlessly support her in her illusory retreat.

Like Dorian, she appeared to have sold her sense of reality in order to be spared its pains. Yet, also like Dorian, who never escaped the awareness of the existence of the picture in the attic bearing all the marks of time passing, she was never unaware of the realities of life, but they were only a source of unavoidable persecution.

As I started to get to grips with some of these issues, there was real progress in her life, although it often remained quite hidden and its connection with the analysis usually disavowed. Her interests widened, she progressed in her career and she developed a deeper aesthetic sense of life, enjoying art and music. This was very much in contrast to her state at an earlier time, when she (rather like Wilde's character, Harry) mocked those who went to exhibitions; she was quite sure they only gave a hypocritical affectation of interest in the art, but in reality wanted only to exhibit themselves. She also became friendly with people who seemed to be more involved in life, able to bear its pains and frustrations and move forward. In time, she became aware that she could no longer maintain the idea that the thing that prevented her moving forward was a dread of breakdown, as she had acquired enough internal resources to make this unlikely and the recognition of this, by both of us, represented a tough, a very important, development.

I want now to bring some material from a time when she was clearly making some important moves. She was more involved in her analysis, and more thoughtful about her life. She had been preoccupied as to whether or not to continue her relationship with her husband or to separate; that is, she wanted to find a way to act, whatever decision she made had real consequences. During this period, there had been increasing awareness of how stuck she was in her illusory world. In particular, she was aware that time was passing and that if she wanted to have a baby, there was only a limited time in which this could be possible. After bringing these issues in a direct and painful way, she missed the next three sessions without any obvious reason.

She started the first session of the following week by mentioning the missed sessions, adding that she could have come but had started to feel that I was not really helping her. She felt "disillusioned" with me. She then made some reference to the previous week, and some sense that something had altered. I suggested to her that she had actually felt that things were improving in the analysis, but that this had brought her into contact with something that she now wanted to evade, namely that analysis was inevitably for her a *disillusioning* process. She was silent for a few minutes and then said, with an air of caution, that she and George (her husband) had bought a painting. (This was something entirely new, she had never bought something of real aesthetic value before.) She went on to describe how she had bought the picture, but the way she did this was quite revealing: it was as if she was excusing herself for having bought it (one had the sense that she felt she was responding to mockery from some observer). George had been left money by an aunt who had died and was anxious to use the money to buy something of worth that would last. She had come across, some time ago, an artist called David Hillbrough, whose pictures she liked. She *happened* to hear that his pictures were being exhibited at a gallery in Camden Town. She *happened* to be window-shopping, and *by chance* passed the gallery without realizing it was the gallery where the pictures were to be exhibited. She went in *for some reason she cannot recall*, and *happened to see* the canvasses stacked on the floor. She did not like some of them but chose one she did, and they bought it. She then said, as an aside, with a hint of acid hostility, "Why should we spend money on such things?" She added that if she and George parted, this would be something that she would have to lose, it would be his.

I was struck by a number of things. First, the atmosphere of sadness and integration, accompanied by a hint of threat. In the to and fro of her session, I suggested to her that buying the painting seemed to represent the capacity to preserve something of value from the life of a dead woman, linked, I suggested, to her dead mother. I suggested that her dead parents might be imagined as providing some resources for her to do the things she liked—going out, looking at paintings, and buying one. I was unsure as how to think of her comment that, if she split up with George, the picture would be his, but suggested that she was aware of losses and that awareness seemed not to be so unbearable today. At the moment she recognized the possibility of having something worthwhile that she could not control, she feared losing it. It was true that if she parted from George she would be ending a relationship that had real value.

I later linked the buying of the picture to the analysis, which was clearly connected to her acquiring new aesthetic depths. I thought that now she was not so "illusioned", she could see me more clearly for who I was. I was no longer seen as just narcissistically exhibiting my ideas for her to admire—they represented my work. She did not like all of it (in the same way that she did not like all the pictures), but recognized in an important way that the analytic work was of real value to her. I thought that, in some way, the loss of her illusions prefigured an awareness that analysis would one day end. (I wondered if she regarded the analytic work as *my* possession, rather than something jointly achieved, like George keeping the picture if they parted, but said nothing about this.) There was a brief pause, and then she said, with humour and real feeling, "My anti-wrinkle cream doesn't work."

I will leave the session there. The principle point I want to make is that my patient, like Dorian, lived in an illusory world and feared that any movement from this state would precipitate the breakdown that had brought her to analysis. She either lived in a timeless world where there was no development, or the only alternative was of being precipitated into a terrifying world in which all the time evaded would suddenly catch up with her, leading to terrors of sudden disintegration and death (which has its parallel in the final scene of Dorian Gray where Dorian moves suddenly from being a man untouched by time to being a "withered wrinkled, and loathsome of visage").

In the session that I reported, it seemed possible for her to move on and develop some capacity to mourn and thus face some of the inevitable vicissitudes of being alive, and it was clear that this brought depth to her life. She regarded this, rightly, as a disillusionment, but *not* a catastrophe. As I have shown there was a terrible tentativeness about this move. She feared the consequences of moving forwards because of the pain and persecuting anxieties it brought to the fore, but further, it was clear that making this move exposed her to a kind of "Harry" figure who looked on and mocked her for exposing herself in this way. She defensively, and transiently, identified with this figure, the continuing more narcissistic side of her character. revealed when she mocked the development with the caustic comment "Why spend money on such things . . .'.

I understood her saying that her anti-wrinkle cream did not work as her acknowledgment that ageing and time passing cannot be evaded, but I thought there was some recognition that there was much to be gained from recognition of the fact.

This material has a clear connection to the issues discussed by Freud in his paper "On transience", where he focuses on the gains to be had from the capacity to mourn. This is central to Klein's description of the depressive position. She described a vital phase in development where the individual acquires the capacity to see the world for what it is: become aware of the impossibility of possessing one's objects and needing to bear the pain of separation from them. This phase of development brings a particular kind of mental pain, a mixture of loss, awareness of separation, and painful feelings of guilt. In order to be able to bear this pain there needs to be already present a secure inner "good object" that is felt to provide support. This ushers the individual into a different world in which there is integration of the ego. The incapacity to bear guilt leads to a dread of looking at any object that is in a damaged state as it stirs up these dreaded feelings.

My point here is that Freud's colleagues, Dorian Gray, and my patient all shared this difficulty. Looking back after some years of the analysis, I was struck by the fact of the absence of any mention of guilt, and it was clear that my patient found such feelings completely intolerable. Once, when they were having a serious conversation about life, her partner described his unhappiness about their lack of a sexual relationship (something he had never done before). My patient's response was to say, "I felt like putting a knife to my throat."

One can therefore read Dorian Gray as a tale of what happens to the character when it surrenders its sense of reality out of a dread of the persecution of ageing. Limitless life in the illusory "retreat" turns out to be life that has little meaning, and in Dorian's case a steady deterioration of character. I mentioned that my patient was severely emotionally deprived in childhood, and this seems to have been true of Dorian, too. We are told in the novel that his father was shot when he was a very young child, and his mother died less than a year later.

Conclusion

In this chapter I have attempted to show how the capacity to mourn, and to bear guilt and loss, are essential to being able to fully apprehend oneself as existing in time; they are different facets of the same problem. If the pain can be borne, the pain inherent in the recognition of the transience of all things, then the subject is instantiated in time and this promotes development. Where this capacity is lacking, existence in time is replaced by the construction of an illusory world where time does not exist. However, life in this illusory world is accompanied by a permanent sense of dread of being exposed to reality, felt as a catastrophic confrontation with a deteriorated and damaged world; a breakdown always threatening and always being evaded. I have illustrated this theme by examination of Wilde's character Dorian Gray, who suffers the catastrophic consequences of life in the illusory world, and also brought material from a patient who showed a capacity, having established some firmer inner foundations, to emerge from this state.

The kind of "psychic retreat" described in this chapter is dominated by a peculiar combination of timelessness, inconsequentiality, and permanent threat. The depressive position brings with it awareness of the self as a person existing in history, existing in time and subject to laws outside one's own control—a world of consequentiality, of causes and effects, essential for the capacity to experience guilt. Awareness of existence in time is both an outcome of the move towards the depressive position and an essential prerequisite for that move .

Dorian Gray can be read from the perspective of what happens to character as a result of the denial of the existence of time, or from the perspective of the consequences of the inability to tolerate guilt. Klein's theory of the depressive position in fact unites these two perspectives, showing their intrinsic dialectical relation.

Perception of time passing, ageing, can be felt to be indistinguishable from perception of damaged objects and so stirs feelings of guilt. The world of cause is also the world of time, the consequential world that cannot be fully apprehended and sustained, where there is no tolerance of guilt. Where this cannot be borne, the result is the timeless quality intrinsic to "life in the retreat". Where there is no time, there is no cause, and so, at depth, life is inconsequential. This confusion of ageing with damaged objects is part of a wider problem that we are all prone to, and which "anti-wrinkle cream" and its various substitutes attempts to resolve for us in an illusory way.

It is a tragic irony that in the wish to stay alive the individual creates a static, dead world. Dorian recognizes that to remain Basil's muse demands that he become a thing and not a person. He says to Basil:

> I am less to you than your Ivory Hermes or your silver Faun. You will like them always. How long will you like me? Till I have my first wrinkle, I suppose. . . . Youth is the only thing worth having. When I find I am growing old I shall kill myself. [p. 34]

Dorian Gray is likened in the novel to Narcissus, who, in the Greek myth we are told, had eternal beauty. However, Narcissus turned away from a pitiable suffering object (Echo), and so from life, and was punished by imprisonment in the dreaded situation that haunts the patients I have described. Narcissus is transfixed to his own image, forced to watch it as it grows old, withers, and dies.

Notes

1. As any great work of literature, *Dorian Gray* can be read from a number of differing perspectives. Green (1979) gives emphasis to a kind of *folie à deux* between the artist Basil and Dorian, where Dorian accepts that he is not free to be a real person but is forced to become the artist's narcissistic ideal object. As Green puts it 'Dorian wishes to remain the unchanging object of adoration, and he is willing to supplant his own reality with another's fantasy I order to attain this goal" (*ibid.*, p. 400).
2. Harry Wotton could be thought of as a vivid representation of a tricky internal object seducing the self into a fascinating perverse. This has been discussed by Gold (1985).
3. The confusion here is of some additional interest. Clearly, there is an important distinction to be made between the "imaginary" as the scene of psychic depth, artistic creativity, the fruit of psychic work, and illusion as shallow wish-fulfilling day-dream, evading psychic work. Here Wilde gives vivid form to a perverse inner scenario: the actress

is seduced into seeing her art as *mere artifice,* and drawn to a world, represented by Dorian/Harry, where there is no reality—all is illusion. See Britton (1995) and Sodre (1999) for further discussion of the crucial distinction between imagination and daydreams.

4. Green (1979) points out that an important aspect of Dorian's relationship with Sybil derives from his appreciation of a more moral, compassionate self, more in touch with reality—"Her trust makes me faithful, her belief makes me good . . ." (p. 91)—and real self, which he sees in her. From this point of view the death of Sybil and his dissociation from it represents one of the last nails in the coffin of his own moral self .
5. For further discussion of unconscious phantasies underlying suicide see Campbell (1997) and Bell (2001).
6. Britton (1999) discusses this further.

References

Bell, D. (2001). Who is doing what to whom: some notes on the internal phenomenology of suicide. *Psychanalytic Psychotherapy, 15*(1): 21–37.

Britton, R. (1995). *Imagination and Belief.* London: Routledge, New Library of Psychoanalysis.

Britton, R. (1999). Primal grief and petrified rage: Rilke's Duino elegies. In: D. Bell (Ed.), *Psychoanalysis and Culture: a Kleinian Perspective.* London: Duckworth.

Campbell, D. (1997). The role of the father in a pre-suicide state. *International Journal of Psychoanalysis,* 76(2): 315–323, also in R. J. Perelberg (Ed.) (1999). *Psychanalytic Understanding of Violence and Suicide.* London: Routledge, New Library of Psychoanalysis.

Freud, S. (1916a). On transience. *S.E., 14*: 303–308. London: Hogarth.

Freud, S. (1917e). Mourning and melancholia. *S.E., 14*: 239–258. London: Hogarth.

Gekle, H. (1986). *Wunsch und Wirklichkeit.* Tubingen: Suhrkamp Verlag, quoted in R. Britton (1998), *Imagination and Belief,* London: Routledge.

Gold, S. (1985). Frankenstein and other monsters. An examination of the concepts of destructive narcissism and perverse relationships between parts of the self as seen in the Gothic novel. *International Review of Psycho-Analysis, 12*: 101–108.

Green, B. A. (1979). The effects of distortion of the self. A study of *The Picture of Dorian Gray. The Annual of Psychoanalysis, 7*: 391–410.

Sodre, I. (1999). Death by daydreaming: Madam Bovary. In D. Bell (Ed.), *Psychoanalysis and Culture: a Kleinian Perspective* (p. 48–63). London: Duckworth [reprinted London: Karnac, 2004].

Wilde, O. (1994). *The Picture of Dorian Gray*. London: Penguin Classics.

CHAPTER FIVE

Regression, curiosity, and the discovery of the object

Rosemary Davies

I started to consider therapeutic regression during a difficult period in the analysis of a patient who I intuited was regressed, and not in a state of what some call "psychic retreat". My researches reminded me of old battles. During the *Controversial Discussions* (King & Steiner, 1991), for example, Ernest Jones recognized it as a potentially explosive concept when he described "this quarrel-provoking word" (p. 323). Neither the adherents of the "nurture regression" school nor those of the "psychic retreat" school resolved the problem. However, Winnicott's linking regression with primary narcissism seemed to me to provide a sound theoretical underpinning for a technique that recognizes the therapeutic value of regression, while avoiding the pitfall of failing to address the destructiveness inherent in the discovery of the object's otherness.

We cannot embark on treatment without consideration of the possibility of the patient's regression. Faced with patients who regress over extended periods of time, rather than the moment-by-moment movements that characterize any analysis, it seems to me that we are somewhat unschooled in our contemporary psychoanalytic culture as to how to work creatively with the patient in this state. Earlier discussions of therapeutic regression got bogged down in issues that amounted to breaches of technique: physical contact with the patient during sessions for example, or in the analyst's self-disclosure. It may have been in response to this that discussion of the centrality of regression in clinical practice diminished. This seems to be a characteristic of psychoanalytic debate in Britain. By contrast, in 1997 the American Psychoanalytic Association debated this topic under the heading: "Therapeutic regression, essential clinical condition or iatrogenic phenomenon?" (Goldberg, 1999).

Therapeutic regression has a distinguished provenance. For Freud, regression was a central concept of psychoanalysis. He outlined a threefold theory of regression: topographical, temporal, and formal (1900a). He argued, for example, that transference, "the most delicate of instruments", was a clinical manifestation of regression (1912b, p. 139). In the context of therapeutic regression, he counselled against "neglect of regression in analytic technique". He was concerned, in his dispute with Jung, that such neglect was tantamount to a dangerous "scientific regression" in itself (1914d, p. 11). In 1936, Kris famously described "regression in the service of the ego". But it is Winnicott (1974) whose work is quintessentially associated with the concept of regression. His work, both theoretical and clinical, privileged the role of "regression to dependence" as he called it.

Bion summarized the various views vividly in his controversial *Cogitations* of 1960 (1992) on why people might think he was not a Kleinian. He wrote, "Winnicott says patients *need* to regress: Melanie Klein says they *must not*: I say they *are* regressed". Bollas (1987) described the clinical phenomenology of regression to ordinary dependence and the "generative regressive process". In accord with Bion, he maintained that the analytic structure "invites regression" so, *ipso facto*, many patients are regressed. Stewart (1992) summarized this position as follows: "regression acts as an ally of therapeutic progress".

Currently, we seem less exercised by overt dispute, but there remains a subtle manifestation of the distaste for the clinical concept of regression. The debate devolves around a false dichotomy between those who argue that regression should be fostered and those who assert that it is tantamount to a psychic withdrawal: the "dialogue of the deaf", as Green (1986) describes it. Britton (1998), for example, considers that regression has appropriately fallen from our psychoanalytic vocabulary. Indeed, Hinshelwood's dictionary of Kleinian terms contains no entry on regression. Britton prefers that we reserve the term regression to describe a retreat into a pathological organization that "reiterates the past and evades the future". However, I think the state of regression only provides an evasion of the future when we fail as analysts to recognize the therapeutic potential of a state of mind that might, more readily than others, give up its unconscious secrets. If we draw on Freud's notion of "formal" regression, we can see how, in the regressed state, primary process, the language of the Unconscious, supersedes secondary process: more primitive modes of expression and representation take the place of the more structured thinking.

Let me set the scene by describing an episode in the treatment of a patient who regressed in analysis.

Stephen

When Stephen started his analysis, I conjectured that he might fall into a depression when analytic work touched some of his defensive structure. His history indicated the probability of some highly charged Oedipal feelings towards a cold, demanding, and humiliating father and a narcissistic mother whom he described as "sweeter than the sweetest cup of tea". However, I did not predict the regression that characterized the first years of treatment.

> In the second session of treatment Stephen reported a dream that became, as early dreams do, a sort of icon in treatment. He dreamt he saw a beached frigate and he knew he could not board it because on board there were diseased people. We saw this as a clear representation of his anxiety about embarking on treatment, where he would have to encounter his own diseased and frightening internal objects. He feared that his characteristic defence of "going numb" would be breached: curiosity about his internal life was consciously experienced as a dangerous venture.
>
> Stephen was often in tears and silent during the early months of treatment. Indeed, he commented, "Every time I think of talking to you I feel like crying." He spoke little and laboriously and in his own words

described himself as "lost in a forest and don't know my way back." He described feeling utterly dropped during breaks and was difficult to pick up again. At the same time, while regressed in analysis, he completed his professional studies and finalized a divorce.

Towards the end of the first year of treatment he told me he felt "intrigued by you". The intrigue was characterized by an erotic element and a very critical censorious element. Despite being analytically informed, he was particularly critical of elements of the setting: he disliked the lack of eye contact and was a stringent critic of any minor lapses in time-keeping. He disliked seeing other patients and worked hard to persuade me that I should alter my schedule. When he told me of these complaints, he would then subside into a terror of the consequences of telling me: an anxiety about the father's cruel injunctions that did not allow for the young boy's curiosity and rivalry.

The silence and hesitancy continued. Then, at the beginning of a session in the second year of his analysis, he reported, "I felt very disturbed yesterday. For a split second when I left my session I thought of following you through the door when I heard you going into your house. I thought I would do something horrible or something horrible would happen to me." He pondered about this frightening image and told me that it reminded him of being a small child when his father would shout brief intense words at him and he would crumble: "a small child reduced to a grain of sand . . . but I know also [remembering a dream he reported the day before] that I can be harmful and damaging."

This was a turning point in his analysis; he was very frightened by the thought of following me. As he walked in his mind through the door into my mind, would he come to grief as he broke through to my house or would I be damaged by his aggressive intrusion? In his associations to this image during the session he remembered "something fundamentally wrong with my parents' relationship". I think he was permitting himself curiosity about his parental objects who faced him with some very murderous feelings. His parents seemed at loggerheads all through his early childhood. But their cruel deception of him was revealed when they had another baby when Stephen was in his early teens.

Stephen gradually emerged from the regressed state and allowed himself to look at the aggression and the libidinal longings inherent in his curiosity about, and discovery of, me as a separate object. He felt assailed by murderous oedipal feelings and reported a dream about the film *Alien* in which a terrifying monster emerges from the body of the heroine. He described his wish to make the danger safe by recruiting the aid of the Ghostbusters.

Regression and the discovery of the object

Emerging from the regression, Stephen was faced with the reality of me as a separate object. I think for many months I had been experienced as little more than a narcissistic extension of his own mind. But, crucially, that recognition revealed the danger of separateness to him. Would he or I be damaged? This clinical moment exemplifies what I regard as one of Freud's central tenets: "The antithesis between subjective and objective does not exist from the first" (1925h, p. 237).

How we conceptualize the nature of the discovery of our own subjectivity and the existence of another seems to me to define our practice. It lies at the centre of the notion of "object usage", where the complexity of the subject placing the object outside the area of their omnipotent control defines a crucial developmental stage.

In this context, Winnicott's writing on primary narcissism and its clinical manifestation in regression proposes a theory of technique that addresses the recognition of our subjectivity: curiosity and the discovery of the object are central. He wrote, "In primary narcissism the environment is holding the individual, and at the same time the individual knows of no environment and is at one with it . . ." (Winnicott, 1974, p. 283).

So, here, the clinical state of regression is equated with primary narcissism. Linking regression with primary narcissism crucially differentiates some writers from others. Balint, a central exponent of therapeutic regression, disagreed with the idea of primary narcissism, preferring to think of primary love. Positing a notion of primary love presupposes the existence of a relationship that is substantially different from the individual who "recognizes no environment".

This is a hotly contested topic theoretically. However, in my view the concept of primary narcissism provides a helpful clinical descriptive tool. It reminds us of a primitive, undifferentiated, wordless state that can characterize the analytic encounter at certain moments. Contemporary evidence points to the recognition of an object from early life that gives credence to the psychoanalytic view of object seeking behaviour over and above instinctual gratification, but it does not entirely account for the detailed phenomenology of the infant's internal state. As Green (2002) puts it

> These observations are behavioral; we still do not know what is going on in the child's mind . . . The baby's reactions to the primary object . . . do not prove that the baby can experience the situation as a separate entity in relationship with another separate entity. [p. 647]

Indeed, from quite another psychoanalytic school, the support for maintenance of the notion of primary narcissism is outlined in Pine's (2004) re-evaluation of Mahler's work on the "symbiotic stage". He argues for the notion of "symbiotic moments". Laplanche and Pontalis (1973) suggest the value of the notion of primary narcissism designating "formative moments". In Gibeault's (2004) discussion of cure in psychoanalysis, he considered primary narcissism and narcissistic regression as the "mainspring of analytic material".

A clinical theory that allows for the concept of primary narcissism demands, then, an explanation of what happens at the point of movement between hallucinated pleasure and dependence on an external object. It is my view that Freud's metapsychology addresses this point. In 1915, Freud's preoccupation with the internal–external axis led him to postulate the presence of aggression inherent in the discovery of the object in the external world. "When, during the state of primary narcissism, the object makes its appearance, the second opposite to loving, namely hating, also attains its development" (Freud, 1915c, p. 139). His well-known assertion that "Hate, as a relation to objects, is older than love" (*ibid.*) is often cited. However, the following elaboration is often not quoted: ". . . It derives from the narcissistic ego's primordial repudiation of the external world with its outpouring of stimuli" (*ibid.*).

Here, Freud's postulation of the hate that is meted out by the narcissistic ego on that which is not "I" is central. Following Freud, Winnicott (1969) asserted the inevitability of aggression in the discovery of the object:

> ... the subject is creating the object in the sense of finding externality itself, and it has to be added that this experience depends on the object's capacity to survive. ... If it is in an analysis that these matters are taking place, then the analyst, the analytic technique, and the analytic setting all come in as surviving or not surviving the patient's destructive attacks. Without the experience of maximum destructiveness ... the subject never places the analyst outside and therefore can never do more than experience a kind of self-analysis, using the analyst as a projection of a part of the self. [p. 714]

The analyst is required to survive the subject's attacks and nurture curiosity and discovery of an object, rather than fostering projective repudiation of the object's otherness. Winnicott makes a crucial distinction between those who argue that projective mechanisms create external reality and his own view that

> projective mechanisms assist in the act of noticing what is there, but are not the reason why the object is there ... orthodox psychoanalytic theory ... tends to think of external reality only in terms of the individual's projective mechanisms. [*ibid.*]

He goes on to argue in the context of the reality principle that destructive feelings are present when the object is discerned. Such a conceptualization indicates the risk of an eruption of destructive phantasies as the patient emerges from the regressed state, placing the analytic object outside of the self. When Stephen recognized a version of me with my own thoughts, my own life, the other side of the door, he or I were indeed in danger.

Primary narcissism is ruptured by the reality principle. The reality principle ushers in a caesura: the subject can no longer live on hallucinated pleasure alone. This is central to our clinical sensitivity. It is in the moment of rupture that we may discover, alongside our patient, the nature of their internal world. The technical challenge will surface when the subject momentarily notes there is something he is not at one with. The analyst is experienced as a representation of the world of "otherness", the "disrupter", so to speak. In the human infant's uniquely lengthy state of dependency, an inevitable loss and rupture occurs: the nursling is faced with a notion of a separate other who gives or withholds the breast. As

Freud (1923b), at his Platonic best, wrote, "the child never gets over the pain of losing its mother's breast".

Absence fuels exploration, and here we see the centrality in psychoanalysis of what drives that curiosity. It is the interruption in relation to the fantasized, inexhaustible breast that initiates her researches: the disruption of the primary narcissistic state.

I conjecture that the regressed state is ruptured, just as the primary narcissistic moments are disrupted, by the recognition of dependence within the analysis: this can be a life or death moment. Rosenfeld (1987) considered:

> When he is faced with the reality of being dependent on the analyst ... [the patient] would prefer to die, to be non-existent, to deny the fact of his birth ... Some of these patients become very depressed and suicidal. [p. 107]

If the regressed state is assumed to be a pathological withdrawal, the default position is often a reliance on interpreting the regression as an attack on the analysis. Following Freud, I am arguing that hate finds expression when the patient emerges from the regression, rather than the regression being the consequence of hatred: the pathological retreat. Destructiveness is definitively present as the subject recognizes the presence of a discrete other. However, this is more usefully assumed to be a representation of self-preservative aggression, not the deadly envious aggression that is assumed in the interpretation of the regressed state as an attack on analytic work, as a pathological withdrawal.

I focus on a particular moment in Stephen's analysis because when he imagined following me into my house, erasing the wall between the analytic space and my home, walking through the door of his mind into mine, I conjecture that this shattered the regression that had characterized the first years of his analysis and faced him with the object beyond his control: it evoked difficult destructive feelings. Stephen required resilience on my part to contend with this psychic movement. This can be achieved without recourse to a technique that ignores the destructiveness inherent in the discovery of the object's otherness. I think this acknowledgement helps the analyst to survive without retaliation: to be an object for another day. In the detail of the clinical moment the analyst becomes a regulator of the regression through the creative, receptive attitude, not through breaches of the setting. It seems to me that we can maintain our analytic stance and adhere to the rule of abstinence, without forgoing a view that regression is a feasible and interpretable state. The analytic stance is not only tested during the regression but also as the patient discovers the hitherto "unknown environment", as Winnicott described it. Stephen confronted this "unknown environment", the "external world with its outpouring of stimuli", the triadic world of oedipal danger, as he imagined crossing the threshold from my consulting room to my home: would he be murdered or murderous, or could a benign, boundary-setting third be summoned? This part of his analysis was indeed characterized by violent imagery.

I think that a more active interpretive technique, aimed at levering Stephen out of his regression in the first months of his analysis, might have precipitated a pathological psychic withdrawal or a premature termination of treatment: the regression would not have been mutative. I felt vindicated by the technical approach to Stephen's state since he managed gradually to recover and recount something of his own history that had been lost to him. Thus, he was enabled to embark on what Loewald (1960) described as one of the aims of psychoanalysis: to restore the patient's sense of historicity by "turning ghosts into ancestors".

Allowing receptivity to the experience of the regression provides a context within which the destructiveness apparent in the discovery of the object can be analysed. Sandler (1993) suggested that there may also be a type of regression, which takes place "in a controlled way" (p. 1104), for the analyst in identification with the patient. Just as the patient is required to give himself up to the associative path, so, perhaps, in identification with the patient, the analyst surrenders to a sort of parallel regression. Similarly, Parsons (2005), in his discussion of formal regression, writes, "Analysts may need to accompany their patients into this domain, relinquishing their grasp on verbal representations and the logical connections between them". Botella and Botella (2005) draw our attention to this "mutual regression . . . an inclusive movement like a double primary identification operating simultaneously in both partners in the session" (p. 105). Sandler likened this sort of mutual regression to certain uses of projective identification as countertransferential markers.

I would like to consider two elements that are illustrated by this technical conceptualization of the analytic attitude to the regressed patient: first, curiosity, and second, what I call a deficit in the language of affect.

Curiosity

In his initial presentation Stephen seemed curious about himself, but his first dream of the beached frigate that could not be boarded for fear of what might be encountered therein indicated an anxiety about investigating too far. However severe the pathology, patients who arrive in the analyst's consulting room retain something of their sense of curiosity. They may just be curious to see a "shrink" in real life, but more probably they are curious about themselves and their predicament. This may vary from conscious recognition of the curiosity about self, to unconscious communication of curiosity about the self in a projected form.

Spillius (1992) reminds us of the centrality of curiosity and the fragility of its sustenance. In her discussion of Steiner's (1994) differentiation of *analyst-* and *patient*-centred interpretations she writes that the *analyst-centred* interpretation suggests a less blaming world

> . . . which fosters curiosity, increased capacity to bear loss and awareness that other people are separate from oneself but have minds fundamentally similar to one's own even though they may have different thoughts.

Emerging from the primary narcissistic shell in the analytic context faces the subject with another who has "different thoughts", not just a projective entity, but also an entity in its own right. Spillius's point alerts us to the "blaming world" the patient often inhabits, which augurs against ordinary curiosity. Sometimes there is a dizzying reflection of this blaming world that envelops both patient and analyst. This can swamp our discourse and perhaps represents a defence against being accused of failing to see the aggression and destructiveness in the patient's material.

Curiosity is a part of a fundamental concept of psychoanalysis: the epistemophilic instinct, the search for knowledge. It is that curiosity which Freud demonstrated throughout his work: curious to hear the "reminiscences" of the hysterics in 1895 and still curious in 1939 in his last papers to discern the scope or limits of psychoanalysis. He assumed a curiosity in us all, and intuited its absence as pathological. Little Hans's enthralling explorations of his body in relation to his mother, father, and newborn sister come to mind as a fine example of the researches of the child. And, of course, it is that epistemophilia which fires our own continuing researches into how the self comes into being. The child's curiosity and explorations interweave with the essential existential plight: where did I come from? That question confronts the child with the parental couple, the primal scene, that union of difference that further alerts her to an environment beyond the not-knowing state of primary narcissism. The primal scene, in all its fantasized forms, provides a screen on to which all sorts of projections and identifications are beamed. Crucially, the child's oedipal curiosity is tempered as desire meets prohibition: curiosity might stimulate dangerous discoveries. You may recall Stephen's realization of "something fundamentally wrong with my parents' relationship" when he reflected on his fear of his own curiosity about me.

In our own practice, anxiety about our technique sometimes augurs against the pursuit of natural curiosity: an overweaning psychoanalytic superego constrains us from taking risks, pursuing unlikely threads. As Pontalis (1981) wrote, it is as if "in contemporary psychoanalysis the individual is captured in another's system" (p. 146). Green (1986) makes a similar point when he writes of:

> imprisonment within the interpretative matrices which translate the unknown into the already known, and are revealed as inappropriate to a mode of relating in which the analyst, by his tolerance of the regressive needs of the patient, might facilitate evolution by declining to fix the experience in a mould which limits his freedom of movement in his psychic functioning. [p. 200]

So, far from the notion of the patient needing to discover the analyst, the as yet unknown environment, the patient can find him or herself trapped in a labyrinthine web of interpretations that can deflect curiosity rather than fostering the free associative process and the mutative analytic endeavour. Current controversy in our interpretive theory reflects Freud's contradictory remarks on, for example, the interpretation of the transference. In his famous papers on technique he writes, "every conflict has to be fought out in the transference" (1912b, p. 104). Then, in the same series, he writes, *"so long as the patient's communications and ideas run on without obstruction, the theme of the transference should be left untouched"* (1913c, p. 139, original italics). Freud's postulation of this unobtrusive transference in my view underpins Winnicott's description of his technique:

> it is only in recent years that I have become able to wait and wait for the natural evolution of the transference arising out of the patient's growing trust in the psychoanalytic technique and setting, and to avoid breaking up this natural process by making interpretations. . . . If only we can wait, the patient arrives at understanding creatively . . . I think I interpret mainly to let the patient know the limits of my understanding. The principle is that it is the patient and only the patient who has the answers. [1969, p. 711]

Sometimes it seems that in our discourse less heed is paid to the excitement of discovery, of being the privileged listener, and more ascribed to preconceived notions of the patient as saboteur. And all this despite deference to Bion's recommendation that we should approach a session "without memory and desire": an injunction that has all but lost its meaning in its wearisome repetition. I think, for example, without joining forces with the self-revelatory Intersubjectivists, that there can be recognition that the analysand's curiosity about his analyst can be interpreted as creative and generative rather than malign and intrusive. This is particularly evident in our attitude to the analysand's enquiries of his analyst. We have to hold the balance between maintaining our neutral, abstinent stance and recognizing that these enquiries reveal much about the patient's internal world via projection and identification. Fisher (2006) describes the inevitable tension between the questions that keep the curiosity alive, spawning more enquiry as compared with answers that foreclose further investigation. He elaborates Bion's thinking on curiosity and postulates a drive for curiosity linked with the establishment of the reality principle. He concludes that

> This [curiosity] surely is the essence of psychoanalysis, the opening up of the analyst to the emotional experience of wanting to know

> the patient, thus making possible by the patient's internalization of this relationship, a wanting, and being able, to know oneself. [p. 1235]

Ogden (2004) specifically addresses the vexed question of the patient's curiosity about the analyst. He argues that in order for the analyst to participate in the process of getting to know the patient, there is also an element of the analyst getting to be known by the patient. He is critical of the view that a patient cannot really "know" his analyst because he does not know what occurs outside the analysis. This idea is flawed according to Ogden because

> it does not sufficiently take into account the fact that, to the extent that the analyst's life experience both within and outside of the analytic setting are significant, they genuinely change who the analyst is. The alteration in his being is an unspoken but felt presence in the analysis. [p. 865]

Stephen's fear of his own curiosity was illustrated when he described being panic-stricken in a bookshop when he caught sight of a woman behind a stack of books. He wanted to see her better but was fearful of being seen to be interested. He associated to a meeting earlier in the day when he described, as he put it, "itching" to say something, but worried that if he did speak his "passion" and "aggression" would be revealed. He agreed that the fear of speaking and the intensity of his curiosity in the wish to see the "bookish" woman behind him, his analyst, exposed him and me to disturbing libidinal and aggressive feelings.

The deficit in the language of affect

To conclude, I would like to consider another problem in analytic technique that sometimes emerges in the treatment of the regressed patient. In the clinical vignette, Stephen was assailed by an intensity of affect as he contemplated the fear of what he might expose himself to if he followed me through the door. In the moment of affective contact, the regressed state was breached. He found himself assailed by powerful affect, and recovering his equilibrium required some hard analytic work. The regressed patient requires us to recognize when to hold off and when to hold forth. When the patient begins to discover the object who has waited, words to describe these affects' return are a crucial vehicle of the analysis. Indeed, recognizing the affective self faces the subject with the notion of the other: "The affect is the epiphany of the other for the subject" (Green, 1999, p. 215). This can present something of a challenge. The wording of affect may leave the patient feeling deficient. The patient's psychic survival may depend on the disavowal of the existence of another who may evoke intolerable feelings. This is constantly under threat in treatment where the analyst is attempting to help the patient to be curious about his internal world and give voice to his affective state.

In my view, this affective dimension is particularly problematic, for the patient is faced with a cruel disjuncture. The analyst's scrupulous attention to his or her affective state in the countertransference contrasts and conflicts with the patient's confused emergence into a world of "otherness" that evokes intense affects. Stephen rather poignantly described this to me when he progressed in treatment and started a relationship with a woman. He returned to a silent state and complained to me of a sense of deficiency, "Both you and she have a language that I cannot use." This can leave him feeling that my emotional fluency, as he sees it, usurps his sense of emotional agency, just as his mother usurped his early experience when she could not let him out of her sight.

Current psychoanalytic theory of technique rightly focuses on the need to pay careful heed to our affective echoes of the patients' material and our interpretive response. It is generally argued that recognizing the affective content within the analytic setting is required before interpretation (e.g., Chused, 1996;[1] Fonagy & Target, 1995, Rosenfeld, 1987). So, while the analyst is schooled to be mindful of his or her affective state and its meaning countertransferentially, the patient may be thoroughly compromised in terms of his affective world. Heimann (1960), in her famous restating of the notion of the countertransference, exhorts us to *sustain* our affective state:

> what distinguishes this relationship from others is not the presence of feelings in one partner, the patient, and their absence in the other, the analyst, but the *degree* of feeling the analyst experiences and the use he makes of these feelings ... [and his ability] to *sustain* his feelings as opposed to discharging them like the patient. [p. 152]

But the patient may be ridding himself of difficult affect that may alert him to the terrifying possibility of the existence of a separate other. So the analyst's capacity to reflect on and verbalize his or her own affective experience, while essential technically, may also intensify the patient's sense of isolation. The analyst who brings these feelings to the patient's attention is underscoring the presence of a mind, the analyst's, which is not the patient's to colonize, but thinks and feels for itself. The patient's observation of the analyst's attempt to deal with these highly charged moments can be mutative, but the patient may also feel threatened by an assumption of a more sophisticated psychic range that he enviously observes in his analyst. McDougall (1989) asks, "How can we give life to those who ask only that we help them keep their prison walls intact—and our emotional reactions to ourselves?" (p. 117).

Conclusion

Regression in the clinical setting is characterized, in my experience, by a highly charged affective encounter difficult to put into words, but crying out for accompaniment. I take the view that this state is, in some senses, a return to an earlier "frozen experience" (Winnicott, 1954). But in another way, it is an entirely new experience for the patient who has taken the risk of letting himself be known in this narcissistically vulnerable state. In this formulation, where there is a dominance of primary process thinking, the potential for change may be greater, for defences are breached and access to unconscious material may be enhanced. I think that if we can maintain a structured analytic stance within our clinical practice, patients in a state of regression can be helped psychically to arrange and rearrange their internal objects in a way that allows them to bear the inevitable realization of separateness. Thus, regression is not the immobilized, atrophied state that critics assume, but an inevitable and mutative aspect of many good enough analyses.

Notes

1. In the treatment of the regressed patients where wording may be problematic, Chused's (1996) conceptualization is particularly apposite. She describes what she calls the *informative experience*: an affective,

sometimes non-verbal communication that lays the foundations for an interpretation. *Informative experiences* are seen as arising out of interactions where the expected reaction is not forthcoming. The resulting emotional dissonance between expectation and experience "informs" and may provide an impetus for psychic change. Chused argues that in this way interpretive communication may be heard without being severely contaminated by superego projections: "learning is always easier from experience than from explanation, which can feel critical to the one-who-didn't-know" (p. 1069).

References

Balint, M. (1968). *The Basic Fault*. London: Tavistock.

Bion, W. (1992). *Cogitations*. London: Karnac.

Bollas, C. (1987). *The Shadow of the Object*. London: Free Association.

Botella, C., & Botella, S. (2005). *The Work of Psychic Figurability*. London: Routledge.

Britton, R. (1998). *Belief and Imagination*. London: Routledge.

Chused, J. (1996). The therapeutic action of psychoanalysis: abstinence and informative experiences. *Journal of the American Psychoanalytic Association, 44*: 1047–1071.

Fisher, J. V. (2006). The emotional experience of K. *International Journal of Psychoanalysis, 87*: 1221–1237.

Fonagy, P., & Target, M. (1995). Understanding the violent patient: the use of the body and the role of the father. *International Journal of Psychoanalysis, 76*: 487–501.

Freud, S. (1900a). *The Interpretation of Dreams*. *S.E., 5*. London: Hogarth.

Freud, S. (1912b). The dynamics of the transference. *S.E., 12*. London: Hogarth.

Freud, S. (1913c). On beginning the treatment. *S.E., 12*. London: Hogarth.

Freud, S. (1914d). *On the History of the Psycho-Analytic Movement*. *S.E., 14*. London: Hogarth.

Freud, S. (1915c). *Instincts and Their Vicissitudes*. *S.E., 14*. London: Hogarth.

Freud, S. (1923b). *The Ego and the Id*. *S.E., 19*. London: Hogarth.

Freud, S. (1925h). On negation. *S.E., 19*. London: Hogarth.

Gibeault, A. (2004). Cure in psychoanalysis. Unpublished discussion paper.

Goldberg, S. (1999). Regression; essential clinical condition or iatrogenic phenomenon. *Journal of the American Psychoanalytic Association, 47*: 1169–1178.
Green, A. (1986). *On Private Madness*. London: Karnac.
Green, A. (1999). *The Fabric of Affect in the Psychoanalytic Discourse*. London: Routledge.
Green, A. (2002). A dual conception of narcissism. *Psychoanalytic Quarterly, 71*: 631–649.
Heimann, P. (1960). On counter-transference. In: M. Tonnesmann (Ed.), *About Children and Children-No-Longer* (pp. 73–79). London: Routledge, 1989.
King, P., & Steiner, R. (1991). *The Freud–Klein Controversies*. London: Routledge.
Kris, E. (1936). The psychology of caricature. *International Journal of Psychoanalysis*, 17: 285–303.
Laplanche, J., & Pontalis, J.-B. (1973). *The Language of Psychoanalysis*. London: Hogarth.
Loewald, H. (1960). On the therapeutic action of psychoanalysis. In: *Papers on Psychoanalysis*. New Haven, CT: Yale University Press.
McDougall, J. (1989). *Theatres of the Body*. London: Free Association.
Ogden, T. (2004). This art of psychoanalysis. *International Journal of Psychoanalysis, 85*: 857–877.
Parsons, M. (2005). Psychoanalysis, art, listening, looking, outwards, inwards. Unpublished paper presented at the British Psychoanalytic Society.
Pine, F. (2004). Mahler's concepts revisited. *Journal of the American Psychoanalytic Association, 52*(2): 511–533.
Pontalis, J.-B. (1981). *Frontiers in Psycho-Analysis*. London: Hogarth.
Rosenfeld, H. (1987). *Impasse and Interpretation*. London: Tavistock.
Sandler, J. (1992). Reflections on the developments of the theory of psychoanalytic technique. *International Journal of Psychoanalysis, 73*: 189–198.
Sandler, J. (1993). On communication from patient to analyst. *International Journal of Psychoanalysis, 74*: 1097–1107.
Spillius, E. (1992). Discussion of Steiner's paper. Presented at UCL Conference.
Steiner, J. (1994). *Psychic Retreats*. London: Routledge.
Stewart, H. (1992). *Psychic Experience and Problems of Psychoanalytic Technique*. London: Routledge.
Winnicott, D. W. (1969). Use of an object. *International Journal of Psychoanalysis, 50*: 711–716.

Winnicott, D. W. (1974). Metapsychological and clinical aspects of regression. In: *Through Paediatrics to Psychoanalysis* (pp. 278–294). London: Hogarth.

CHAPTER SIX

The Aztecs, Masada, and the compulsion to repeat

Gregorio Kohon

> "... foreseeing is part of knowing. And historians are constantly foreseeing. If only retrospectively"
>
> (Eric Hobsbawm)

> "There is no alternative: one must impose a meaning on what perhaps has none ..."
>
> (Elie Wiesel)

I

In his accomplished short history of the twentieth century, *The Age of Extremes* (1994), Eric Hobsbawm writes:

> ... it is not the purpose of this book to tell the story of the period which is its subject ... My object is to understand and explain *why* things turned out the way they did, and how they hang together. [p. 3]

Hobsbawm then proceeds to argue that the major task of a historian is not to judge but "... to understand even what we can least comprehend". And a few lines later, faced with the need to understand "the Nazi era in German history and to fit it into its historical context", he confesses, "no one who has lived through this extraordinary century is likely to abstain from judgement. It is understanding that comes hard" (p. 5).

This concern for understanding and the struggle to achieve it bring together history and psychoanalysis. In Michel de Certeau's understanding of history, this discipline would be formed and constituted into a "science of the Other". He called this a *heterology*, which necessarily incorporates the fundamental discoveries of Freudian psychoanalysis: the return of the repressed, the presence of the imaginary in the rational, the insistence of the unconscious in science, the unavoidable mixture of fact and fiction in historical reconstruction (de Certeau, 1986).

Hobsbawm stresses the search for reasons, for the causes of historical facts: *Why*, he asks, have things turned out the way they have? A historian, rightly and of necessity, overlooks the distinction introduced by Dilthey (1894, quoted in Klauber, 1981, p. 185) between the *understanding* of an event and the *explanation* of such an event. Similarly, it is at the heart of the Freudian enterprise that, even though meaning should not be confused with cause, they are undeniably linked to one another. It is only because we search for historical causes that we can explain a psychical event, that we are in a position to find meaning in it. The reverse is also true: because a psychical event has a meaning, we are motivated to look for its causes. The parallel does not end there: in both disciplines, when we talk of cause or causes, we are also implicitly referring to the overdetermination of all events, whether historical or psychological.

This view of history as a hermeneutic discipline has been put forward by a large number of historians: among many others, Croce, Collingwood, Cohen, Hegel, and, closer to psychoanalysis, Peter Gay. They have all stressed the role of values and of critical reflection, and above all the need for interpretation. Furthermore, Peter Gay, in his early book *Style in History* (1974), has argued against the polarity of private subjectivity *versus* professional objectivity, which he believes is an impossible demand made upon the work of the historian. Taking the texts of authors like Gibbon, Ranke, Macaulay, and Burckhardt as examples, Gay believes that it was through their subjectivity that these authors had arrived at objective historical knowledge. Passion, even prejudice, Gay passionately argues, can provide access to insights (Gay, 1974). And in a later book, *Freud, Jews and other Germans—Masters and Victims in Modernist Culture* (1978), he declares that historians are exposed to the same risks that psychoanalysts are in their practice. They can also find their work distorted by feelings of affection or aversion towards their subjects, a phenomenon that psychoanalysts call countertransference. In particular, Gay says, "The writing of German history is laden with, mainly unexamined, counter-transferences" (*ibid.*, p. ix). Like the analyst, the historian is also forced to confront and examine his countertransference, so as to arrive at his interpretations.

Given the peculiarities of both disciplines, history and psychoanalysis are faced with a similar theoretical challenge: how to validate a knowledge that operates through the very subjectivity of the practitioner of the science. Historians, as much as analysts, partly infer their hypotheses and their conclusions from indirect evidence. Isaiah Berlin (1996) has argued that history does not consist of a mere recital of facts belonging to the past. These facts need to be set in the "continuous, rich, full texture of 'real life'—the intersubjective, directly recognisable continuum of experience" (p. 26). This kind of understanding, Berlin says, can only be obtained by an "imaginative insight" (p. 25), a complete subjective act of imagination. Thus, historians interpret their evidence moving beyond and behind the appearance of the events. And, in order to be able to do this, the historian must possess "imaginative power of a high degree, such as artists and, in particular, novelists require" (Berlin, 1997, p. 353). It is absurd to presume or to assume that this activity provides us with absolute, objective knowledge, which could be measured by the parameters of other sciences. Nevertheless, this should not prevent us from arguing that in history as well as in psychoanalysis, we are dealing with truth—a truth that in both cases we may call *historical*. It is historical as opposed to being "precise", in the mode of the natural sciences. Historians (and the same can be said of anthropologists and psychoanalysts) deal with an "unparalleled knowledge of the varieties of human social experience" (Hobsbawm, 1997, p. 52). And yet none of them can demonstrate the truth of what they are saying by "reference to theories or systems of knowledge, except to some inconsiderable degree"

(Berlin, 1996, p. 33, fn). Truth here is based—like in imaginative literature—on subjective, intuitive experience (Berlin, 1997, p. 165).

Finally, and before moving on to the subject of this paper, there is one more important element that brings psychoanalysis and history together. That is the issue of time.

Time will be used in the context of this presentation, not in its strictly chronological sense, applicable to a passing diachronic instant, but as a logical structure in the human mind. Logical time is the adjudicator of human understanding. "Logical" implies that it is not linear, which is what characterizes chronological time. In psychoanalysis, the concept of logical time necessarily encompasses notions like the a-temporality of the unconscious (as described by Freud), its negativity (as postulated by André Green, 1993), as well as that of *Nachträglichkeit*. It is logical in the same sense that one speaks of the logic of the unconscious, or the logic of the transference; it is an *exploded* logic where time passes and does not pass at one and the same time. Events do not end at a certain given chronological point; they continue to exist staying the same, while going on transforming themselves.

A word on *Nachträglichkeit*, a psychoanalytic concept that has been wrongly translated as "deferred action". In French, the term used is *aprés-coup*. Thomä preferred "retrospective attribution" (Thomä, 1989); Arnold Modell calls it a "retranscription of memory" (Modell, 1990); Borch-Jacobsen speaks of "aftermath-effect" (Borch-Jacobsen, 1996); Laplanche has suggested the term "afterwardsness" (in Fletcher & Stanton, 1992). The definition given by Laplanche and Pontalis is as follows:

> Term frequently used by Freud in connection with his view of psychical temporality and causality: experiences, impressions and memory-traces may be revised at a later date to fit in with fresh experiences or with the attainment of a new stage of development. They may in that event be endowed not only with a new meaning but also with psychical effectiveness.[1] [p. 111]

When we deal with the history of an individual in the psychoanalytic situation, *Nachträglichkeit* makes it impossible to merely explain the present through the events of the past. Nothing can be reduced to a linear sequence of events; logical time transforms the subject's past into a historicization of his present. The same might apply to human history in general, which cannot be explained as a result of a simple cause and effect deterministic development. This is not just the opinion of a biased psychoanalyst; in fact, it is also the view of certain historians, notably Benedetto Croce, who believed that history must be a re-enactment of past experiences. In the words of the British historian Robert Collingwood: "the past, so far as is historically known, survives in the present" (Collingwood, 1946, quoted in Klauber, 1981).

I will now refer to two different examples taken from general history.

II

The first example is from the history of the conquest of Mexico. After fighting the Muslims for eight centuries, and ordering the expulsion of Arabs and Jews from their land, the Spaniards found new places to conquer and new people to whom they could preach the Gospel. After the discovery of the New Continent in 1492, the Spanish Crown did not have a *Cronista* (chronicler) documenting *la Conquista de las Indias* until 1620, when the official post was then created. The *Crónicas* were politically biased accounts, as distorted as the previously unofficial reports. Naturally, there had been a fair number of chronicles before that date, some independent, but the majority of them were reports to the Spanish Crown. The history of this period was told by three different kinds of writers: (1) the *Conquistadores* themselves, who were mainly interested in justifying their violent actions; (2) the descendants of the Amerindians, who attempted to show the past glories of their race; and (3) the missionaries, who made a special effort to demonstrate the advantages the natives had gained from the introduction of the Gospel. Among the early accounts of the conquest of Mexico, the most famous authors are: Hernán Cortés, whose *Cartas-Relación* (Letters) were sent to the Spanish monarchs during a period stretching between 1519 to 1526, and published very soon after; Bernal Díaz del Castillo, a foot-soldier who accompanied Cortés; Francisco López de Gomara, whose *Historia General de las Indias* fell under the censorship of the Crown in 1553; and possibly the two most important writers, Sahagún, with his *Código Florentino: Historia General de la Cosas de*

Nueva España, and Fray Diego Durán with the *Historia de las Indias de Nueva España y Islas de Tierra Firma*.

The history of the Conquest of Mexico has been of special interest to historians, and the writings on this period are very extensive. Based on the accounts written by the Conquistadores, the descendants of the Amerindians, and the missionaries, the facts concerning the conquest of the Aztecs by Hernán Cortés are fairly well known and more or less agreed on by historians. The main events can be summarized as follows.

Hernán Cortés lands on the Mexican coast in 1519. His expedition, consisting of several hundred men, disembarks in Veracruz. Cortés, against the express orders of the Governor of Cuba (who wanted him to return there), starts his march to the interior after hearing of the existence of a rich and most impressive imperial city. A very able manipulator and warrior, Cortés pushes his way through many lands, their people succumbing either to his deceitful promises or to his military strategies. He wins over to his cause (which became, ultimately, the conquest of the Aztecs) indigenous tribes like the Tlaxcaltecs, powerful enemies of the Aztecs who then become Cortés' best allies. Cortés and his men finally arrive at Tenochtitlan-Tlatelolco, where he is well received. After a while he decides to take prisoner Moctezoma, the "Great Speaker", the Emperor of the Aztecs. In the meantime, the Governor of Cuba sends a large group of armed men to capture the rebel Cortés; the *conquistador* sets out to meet them, leaving some of his soldiers to guard Moctezoma. Cortés defeats the rival Spanish army, and wins the survivors over to his cause. In the meantime, during a religious festival, the men he leaves behind in Tenochtitlan-Tlatelolco massacre a group of Aztecs. War was then declared, and Moctezoma dies, probably assassinated by his jailers. Cortés arrives, and after being reunited with his men, decides they should secretly escape during the night. They are discovered and half of them are killed in the ensuing battle. This is the famous *Noche Triste*. Cortés manages to escape to Tlaxcala. Once reorganized, he returns with what is left of his fragmented army, and with the help of his new allies, the Tlaxcaltecs, lays siege to the capital. After several months, Mexico City falls and the Aztec "Empire" collapses.

This is a simplified sketch of the events, which nevertheless gives the flavour of *what* happened. For all its complications, it comes down to this: a small bunch of Spanish adventurers led by Hernán Cortés, without any prior knowledge of the territory, the culture, and languages of its people, was able to claim victory over several hundred thousand men and women, the subjects of a powerful empire, after merely two years of fighting. To the people who study this period, the Mexican conquest remains one of the most enigmatic historical events. It is difficult to know *how* it happened, let alone *why* it happened.

One reason given by historians is the strange behaviour of Moctezoma: the sovereign did not offer much resistance, allowed Cortés to take him prisoner and showed himself to be hesitant and perhaps afraid. He indicated to the invaders that he knew they were coming, and offered his kingdom as a present, while at the same time asking the Spaniards to return to Spain. In the meantime, his sole concern seemed to have been to avoid any confrontation. Thus, he surrendered his enormous power. His behaviour, sometimes seen as an act of cowardice, and at times as the result of madness, continues to baffle historians. It is indeed perplexing, and appears inexplicable.

As soon as Moctezoma died, the Aztecs declared war. It should be noted that the Aztec kingdom was made up of many different groups of people, some of whom had previously been conquered, colonized, and abused by the Aztecs. The idea of a powerful Aztec empire living in harmony is an illusion created by the invaders and their chronicles; in the words of a historian, the different groups of people were "held together by the tension of mutual repulsion" (Clendinnen, 1991). The Aztecs were certainly powerful, but they did not constitute an "empire". The divisions among them and the enmities with other groups of Indians were well exploited by Cortés, who offered himself as a lesser evil than the Aztec rulers. Toward the end of the war, Cortés was heading a powerful "Spanish" army mainly composed of Tlaxcaltecs and other Indian allies.

Moctezoma's peculiar behaviour and the divisions among the Indians are, then, two of the known factors that would help to explain Cortés's victory. There were two other important reasons that contributed to the defeat of the Indians: (1) the difference in weapons used by the two sides, and (2) what would be called today bacteriological warfare, with the introduction by the invaders of smallpox, which ravaged the local population.

And yet, these reasons do not seem to convince historians as to why the Aztecs were defeated. There are many questions left unanswered, and the very interpretation of the events seems to provoke great disagreement. One instance is the characterization of the two main protagonists presented in the diverse historical accounts: Moctezoma, on the one hand, as the naïve, indecisive, cowardly Mexican ruler; Cortés, on the other, as the ruthless, pragmatic, and intelligent Spanish conqueror. This seems to be an oversimplification. The differences, contrasts, and misunderstandings between those two warriors run deeper and are more complicated.

Inga Clendinnen (1991) says that:

> Cortés interpreted Moctezoma's first "gifts" as gestures of submission or naïve attempts at bribery. But Moctezoma, like other Amerindian leaders, communicated at least as much by the splendour and status of his emissaries, their gestures and above all their gifts, as by the nuances of their most conventionalized speech. [p. 70]

The gifts that Moctezoma gave to the invading army were signs of his power and his wealth, not tokens of his submission. It was a splendid and sublime gesture, given in "arrogant humility"; the Spaniards could not understand this, and they "lacked both the wit and the means to reply". Similarly, Cortés could not possibly understand, nor could he accept, Moctezoma's simple argument that while the Christian religion might have been best for the Spaniards, the Aztec religion was best for his own people. Clendinnen makes it very clear that the few identifiable confusions ("only a fraction of the whole", she says) between the two parties were important and went both ways:

> For example, Cortés, intent on conveying innocent curiosity, honesty, and flattery, repeatedly informed the Mexican ambassadors that he wished to come to Tenochtitlan "to look upon Moctezoma's face". That determination addressed to a man whose mana was such that none could look upon his face save selected blood kin must have seemed marvellously mysterious, and very possibly sinister. [p. 71]

Contemporary historians have departed from the traditional, formerly established modes of their discipline. Most of them are now concerned with reading the signs of a particular period, or of a given culture, in order to discover "the social dimension of thought"; they endeavour to find meaning in the historical documents by relating them to "the surrounding world of significance" (Darnton, 1984, p. 14).

Reading the signs: this tests, shifts, and elucidates the boundaries of meaning. Wars, for example, are "not quite as cultural as cricket" (Clendinnen, 1991), but they are none the less fought by signs and bound by rules that determine their meaning. The two sides in Mexico were engaged in a war fought by misreading each other's signs; they were not playing the same sport. It was as if one side was playing according to the rules of soccer, the other according to the rules of rugby.

The misunderstandings thus multiplied. The Spaniards imposed their own view of how wars were supposed to be waged: any means justified the end, which was to win, to be effective. For the Aztecs, wars required an act of cooperation; rules were essential for battles to be fought. Two different concepts of war confronted each other; two different conceptions of truth were at war. For the Aztecs, war was a genuine, unquestionable moral conflict. They did not conceive of the possibility of lying and betraying the enemy; for them, wars were accompanied by rules and treaties, which were there to be respected. On the other hand, while claiming to be motivated by moral virtues, the Spaniards fought their war as they had always done in the past: they lied, cheated, betrayed, and, after transforming the Indians into their hated enemy, they starved, mutilated, and murdered them. Such was their mercy. The Dominican missionary Bartolomé de las Casas, in his famous *History of Indies*, written in the first half of the sixteenth century, declares:

> Not once but many times a Spaniard would ask an Indian if he was a Christian, and the Indian would reply: "Yes, sir, I am a bit Christian because I have learned to lie a bit; another day I will lie big, and I will be a big Christian." [*Historia* III, 145, quoted in Todorov, 1982, p. 90p

Like the Crusaders, the *Conquistadores* wanted to believe they were fighting as Christians against barbarians and infidels. In the process they became indifferent to human suffering, and were driven only by greed. Neither of the two parties was able to translate the other's experience: an alien world that did not make sense failed to be translated into something familiar, knowable, digestible, or (as the *Conquistadores* preferred to say) "natural". Both parties attempted to perceive events as an act (seen from their own experience and practices) that substituted another act (which appeared to their eyes as completely incomprehensible). This was not done by analogy or metaphor, where comparisons might help to negotiate the gap between self and others. The other's behaviour could not be recognized, there was no common ground for any possible recognition; the oceans separating the two cultures could not be crossed. The other was an alien, he was not human. (For other examples of this kind of misunderstandings between the conqueror and the conquered see Sahlins, 1985.) The main thrust of de las Casas' arguments to defend the rights of the Amerindian peoples was based on their being "men like us". This "not being human" created the possibility for the Spaniards to treat the Indians with such cruelty and sadism, like the Crusaders with the Arabs, whom they roasted and ate to prove that they were animals (Segal, 1991, p. 147), and the Nazis with the Jews, whom they called *Untermensch*—subhumans.

What provoked such profound misunderstanding in the reading of the signs on part of the Amerindians? What made Moctezoma such a blind and willing accomplice of his own and his people's destruction?

The different notions of *fate* and *time* that the Mexicans had, compared to the Spaniards, partly account for the contrasting attitudes towards the final battles, the methods employed to fight them, the strategies to approach the enemy, and the explanation offered for the results. Time, for the Aztecs, was multi-dimensional and eternally recurrent; time moved in very complicated ways.[2] Their calendar consisted of thirteen months of twenty days each; each day did not proceed from the previous one; each had its own character, indicated by its name derived from the time counts. The character of the day was defined in terms of it being propitious or unlucky; this meant that the events occurring, and the actions performed on that day, were also either propitious or unlucky. This applied even more to the persons born on that day. As Todorov (1982) says, "To know someone's birthday [was] to know his fate" (p. 63).

The Aztecs were great believers in the power of divination; the interpretation of omens was an essential part of their lives. For example, wars were sacred and nobody could know the outcome of a battle. Nevertheless, since outcomes were predestined by the gods it became imperative to look for the signs indicating which side was going to win, which was going to be defeated. If the enemy took one's banner, for example, this was not an act that hurt one's pride: it was a sure sign that defeat was certain (Clendinnen, 1991). Nothing happened at random. All events outside the ordinarily expected became an unlucky omen. This could vary from a prisoner behaving unexpectedly (e.g., crying), to a mouse running through the temple during a religious ceremony. If the event was not expected, it should not have occurred, and if it did occur, that certainly meant bad news. Everything was based on divination; the soothsayer was consulted—whether it was at the individual and personal level of decision-making or at the level of great decisions concerning the state:

The whole history of the Aztecs, as it is narrated in their own chronicles, consists of realizations of anterior prophecies, as if the event could not occur unless it has been previously announced: departure from a place of origin, choice of a new settlement, victory or defeat. Here, only what has already been Word can become Act (Todorov, 1982, p. 66).

For the Aztecs, fate could not be fought against, nor could it be defeated or avoided: everything was anticipated, there was a timely order in the world that could not be transgressed. While the Spaniards concentrated on thinking of practical strategies to defeat their powerful enemy, the Aztecs were busy deciding how they were going to know what to do: they consulted the soothsayers, read the signs, divined the oracles. This time, they were not very lucky:

> They asked the gods to grant them their favours and the victory against the Spaniards and their own enemies. But it must have been too late, for they had no further answer from their oracles; then they regarded the gods as mute or as dead. [Duran, III, 77, quoted in Todorov, 1982, p. 62]

In considering the history, not only of the conquest of the Aztecs, but also of other peoples of Latin America (the Mayas, the Incas, or the Carib Tainos), something remarkable and dramatic comes into view. Every Amerindian account, every narrative of colonization, starts with the "omens" that had previously announced the arrival of the Spaniards, of their weapons, of their horses. This is the crucial point of psychoanalytic interest: for the indigenous people of America, any event that had been experienced as strange or unexpected *before* the arrival of the Spaniards, was then identified *retrospectively* as a message from the gods. The actions of humans were almost irrelevant to the outcome of a history that had already been written. Todorov (1982) declares: "The Aztecs perceive the conquest—i.e., the defeat—and at the same time mentally overcome it by inscribing it within a history conceived according to their requirements" (p. 74).

The Amerindians revised past events at a later date, through the process that Freud described as taking place *nachträlich*. The "omens" were given significance through the events that followed them. Pre-Columbian people made sense of the present in this way, and this allowed them an easier acceptance of whatever events happened to take place, since they had been "announced" in the past. The arrival of the Spaniards was traumatic from the very beginning because the conquerors were so different, so untranslatable, that the only way to perceive these invaders was to see them as "gods", literally.

Moctezoma "knew" of the return of their god, Quetzalcoatl, the Feathered Serpent, the patron of the priests, the inventor of the calendar and of books, of learning and writing, as much as the symbol of death and resurrection. The legend was that, after having been defeated by another god (Tezcatlipoca), Quetzalcoatl had gone on a sea voyage to the east. The Aztecs believed that one day he would return, and Cortés fitted the bill quite well: his beard, his mask, his conical hat, all gave him away. The Aztecs were able to describe the arrival of their god, but they had no words, or concepts, to describe the actions of *this* god. It was only afterwards, once the Indians had been conquered, that the Spaniards created and made available for the Aztecs new descriptions for the invaders' actions. This sudden irruption, then, was the unassimilated experience to which it proved impossible to attribute meaning. To make any sense of the original trauma (the primary perception of the arrival of the invaders and, above all, the fact that their actions appeared so incomprehensible), the chronicles written by the Indians (whether Aztec, Mayan, or other) understood the defeat (the secondary event) through a retroactive reference to the "omens". It is only then that the primary trauma gained a new meaning, proving its psychical effectiveness: "The fact that defeat was suffered", says Clendinnen (1991), "declares it to have been inevitable" (p. 84).

Human beings construct a self for themselves, an image of who they are, through the memory of what they have been in the past. We tell stories to ourselves and to others about our past, which contribute to create a self in the present based on what we call "memories". It is already well accepted, and sufficiently proven, that we cannot remember things from the past in an accurate way. Memory is not a reliable source of exact information. But the question is even more troublesome: psychoanalysis has taught us that we cannot easily determine *what we have done or what has been done to us,* beyond what we remember or forget. As we constantly change what we have been in the past, we hope to change who we are in the present. If the stories told by us are always the same, we are caught in an unchangeable past that repeats itself; we are *self-institutionalized,* inhabiting a self which is tyrannized by its past (see Kohon, 1999, 2005). In this case, there is no possible creativity in our lives: change cannot be imagined.

Let us recapitulate at this point.

1 We have seen how the Aztecs' myth of the return of their God, *Quetzalcoatl,* determined their perception of the *Conquistador* Cortés as their god. In this case, a myth from the past determined a behaviour in the future.
2. At the same time, the explanations given *a posteriori* by the Amerindian chronicles were justified through the retroactive references to the so-called "omens" offered to them by their gods.

It would be absurd to reduce all explanations for historical events to psychological causes. The consideration of psychological reasons in the understanding of historical facts is only meant *to add* to the complexity of the picture, not to reduce or simplify it. And yet, from a psychological point of view, we might now ask: to what degree did the "omens" which determined the future of the native peoples all over the Americas at the time of the Conquest, also become constitutive of the different "presents" in later periods of their history, after the *Conquista*? Generation after generation of the descendants of those American Indian nations, from Mexico to Tierra del Fuego, the poor, the dispossessed, the forsaken peasants, the last surviving tribes of Indians, to what extent are they living still, today, under the weight of a past that continues to be an important factor in the determination of their destiny?

Is there something in the mentality of the present victims in Latin America that, proceeding from the past, is forever compelling them to be conquered once and again, to be repeatedly raped and abused, prostituted and betrayed?

III

Masada, the fortress built at the top of a rocky mountain of the Judean desert, was the scene of a dramatic episode in the history of the Jewish people.[3] It was the last stronghold (the previous two were Herodium and Machareus) held by the Jews against the Romans, three years after the Roman conquest of the rest of Judea and the destruction of the second Temple of Jerusalem by Titus in 70 AD. Between nine hundred and a thousand children, women, and men (all members of the sect known as the Zealots) died at the top of this 1,300-foot mountain overlooking the Dead Sea, after a lengthy siege laid by the Romans. The Zealots were a sect that rejected Roman rule; their political intransigence was based on theological and messianic beliefs, which brought them into opposition, not only against the Romans themselves, but also their fellow Jews who had accepted the Roman dominion. The Romans referred to them as the Sicarii. They were responsible for starting and taking a leading part in the disastrous war against the Romans. In their fanaticism, they engaged in intensive strife with other segments of the Jewish population, massacring their political opponents on more than one occasion. They also practised different forms of terrorism; for example, murdering Jewish collaborators in the crowds at festival times. (The name of the Zealots comes from the story of Phinehas in the Book of Numbers. Phinehas had saved Israel from the plague by killing a wicked man and his wife with a javelin; he was said to have been "zealous for his God" (Numbers 25: 7–15).)

When the Romans finally demolished the walls of the fortress, the leader of the Jewish dissidents convinced his men to kill their own families, then each other, and finally themselves. The only source for this bit of history (based on the account given by two surviving women and their five children, who had hidden in the conduits that brought underground drinking-water) was offered later by Josephus, one of the former commanders of the Jewish revolt. In his book *The Jewish War*, he described the shock inflicted on the Romans when, upon entering Masada, they discovered that all those who had taken refuge were dead. The exultation of victory anticipated by the Romans was thus denied to them.

Historians have questioned Josephus's version: after being involved in the rebellion against the Romans, he was later employed by the Romans themselves to write the story. In Jewish eyes, he was a traitor for deserting to Rome after the siege of Jotapata. In fact, there are a number of reasons to doubt his account. The most prominent, of course, is the fact that suicide is expressly prohibited in Rabbinical Law. In the first place, the Torah itself has been given to the Jews as a *guide for life*, never as a preparation for death. The prohibition extends not only to suicide, but also to any form of self-mutilation. In his account of the oratory given to his followers by Eleazer ben Ya'ir (the leader of the Zealots), Josephus used arguments that came mainly from the Greek tradition, rather than the Jewish one. For example, in Josephus's account, Ya'ir claims that God had given the Jews freedom to choose their own death; that God had condemned the Jews to extinction; that the war was irremediably lost; that there was no hope for a successful outcome. Furthermore, Ya'ir argues that life, rather than death, is the real calamity. All these arguments have no place in the Jewish tradition; however, they can easily be linked to Greek determinism and Stoic fatalism.

On the other hand, Josephus was acknowledged as a scholar, he understood the Jewish people, and the accuracy of his observations has been supported (save minor errors) by archaeological excavations. There are also some clearly Jewish elements in Josephus's reconstruction of the speech (presumably) given by Ya'ir. For example, the references to the land given by God; the faith in the only God of the Jews; the conception of the Jews as the chosen people, favoured by God; the reference to the strong family attachments of the Jews; and finally, the trust in God and the belief in immortality. Although these characteristics would not favour the idea of a Jewish group committing collective suicide, there remains the uncertainty of not being able to assess the impact of the speech of a charismatic leader on his fanatical followers. It is not rare for fanatical groups—whether political radicals or religious extremists, in any case groups whose life is rigidly structured—to think (not necessarily *consciously*) of suicide as the supreme form of loyalty.

Whether mythical or real, Masada has come to play an important role in the history of the Jews. This has been further established by the excavation of the site, from 1963 to 1965, carried out by archaeologists and thousands of volunteers from all over the world, under the direction of Professor Yigael Yadin (1966). Among the many spectacular treasures discovered during the excavation were Herod's villa (which contained rooms with frescoes and mosaic floors), and his ceremonial palace; stone balls hurled at the defenders by the Romans; the biggest collection of Jewish and Roman coins of the period; the ruins of the earliest known synagogue and *mikveh* (ritual bath); the earliest known manuscript of Ecclesiastes, etc. With all this in view, Masada became more alive than ever before in the Jewish imagination. Entire skeletons, personal belongings, armour-scales, and arrows were mute witnesses of the siege. Storerooms had been left chock-full to prove to the Romans that the mass suicide had not been dictated by hunger. All this, and above all the ostraca[4] found (which included what seemed to be the lots cast by the last ten survivors to determine who was to kill the other nine and then himself) had a very special impact on the contemporary collective memory of the Jewish people (Johnson, 1987).

Memory (whether personal or political, individual or belonging to a particular group) is always the result of a collective process. Halbwachs (1980), through his work on *memoire collective*, established that memory is an organic part of social life; memory, structured through social frameworks, is the living experience of social groups.[5] In Judaism, the injunction to every Jew to remember is unconditional. *Zakhor* is the Hebrew word for *remember!*, a biblical commandment for the Jews. In the words of a Jewish historian: "As Israel is enjoined to remember, so is it adjured not to forget. Both imperatives have resounded with enduring effect among the Jews since biblical times" (Yerushalmi, 1982).

And yet, as extraordinary as it may sound, Masada seemed to have been generally "forgotten" for eighteen centuries by the Jews. Why was the incident never mentioned in the rabbinical literature? And why was it later "remembered"?

Masada seemed to have disappeared from Jewish historical accounts until the nineteenth century, with the rise of European Zionism. For the Zionist settlers of the early twentieth century in Palestine, Masada became a significant symbol: it stood as an example of a heroic war of national liberation. Zerubavel (1994) explains it thus:

> For the Jewish settlers in Palestine and especially for their native sons and daughters, the first generation of "New Hebrews", "Masada" was not simply a geographical site, nor was it merely an episode from Antiquity. It represented a highly symbolic event that captured the essence of the authentic national spirit and helped define their own historical mission as the direct followers of the ancient Hebrews. The Masada episode, marking the end of the Jewish Revolt against the Romans, embodied the spirit of active heroism, love of freedom, and national dignity that, according to Zionist collective memory, had disappeared during the prolonged period of Jewish life in Exile. [p. 75]

After its "discovery", Masada became a patriotic symbol, a lesson to be learned by the Jews settling in Palestine. As might be expected, the story needed to be trimmed and reshaped if it were to fulfil the first settlers' educational aims. Above all, what needed to be stressed was the courage of the Zealots in defending their position on top of the rocky mountain. Although there is no historical record of an actual fight having taken place between the Jews and the Romans, a battle was assumed to have taken place. People were told that the group of Jews besieged by the Romans had fought "to the very end", falling "in battle". Political history was in the process of being written; this is still now one of the popular versions of the events at Masada. It is important to note that:

> The Israeli national ritual constructed round the siege of Masada does not depend on the historically verifiable truth of the patriotic

> legend learned by Israeli schoolchildren and visiting foreigners, and is therefore not seriously affected by the justifiable scepticism of historians specializing in the history of Roman Palestine. [Hobsbawm, 1997, p. 275]

The "patriotic" death of those who died did not include any reference to the issue of mass suicide. A powerful contrast characterized the Zionist revision: they glorified the Jews of Antiquity, and aggressively disapproved of what they saw as the submissive mentality of the Jews of the Diaspora. In excluding any mention of a supposed collective suicide by the Zealots at Masada, the issue of the religious attitude towards the events was conveniently avoided. The Zionists did not have to ask: was this event truly an historical example of *Kiddush ha-Shem*—martyrdom, the only exception in the Jewish religion to the strict rule that forbids Jews to kill themselves? Naturally, Zionism was interested in presenting a positive "fighting" image that would help its cause: to commit suicide was not on the activists' agenda.

Then, there was Hitler. The news reaching Palestine at the time made the settlers more determined than ever to strengthen their resolve to fight for a land of their own. The destiny of the Jews in the European Diaspora only confirmed the settlers' view that Palestine was the only possible safe homeland. Submissiveness was not a choice. They were ready to fight to the end, and die (if necessary) a dignified death. Masada was then reinforced as a model and a symbol of national revival in Israel. Exile only meant being destroyed. It is quite remarkable that for almost twenty years, until the early 1960s, references to the persecution and extermination of the Jews by the Nazis were more or less eschewed in Israel. Schools devoted little attention to the main events, giving priority to the ghetto uprisings and, in general, to the acts of heroism on the part of the Jewish partisans. The symbol of Masada was affirmed as representing a fight "to the end". The ghetto fighters themselves spoke of their uprising as the "Masada of Warsaw". At this point, Masada became associated with a "myth of Renewal", which can be summarized thus: the Jews of Palestine were able to survive *because* of the events that had taken place at Masada; the Jews of today have continued to survive because the Zealots of Antiquity died fighting against the Romans.

Then something else happened. Two major events occurred which set off important changes in the culture of Israeli society (Zerubavel, 1994). One was the trial of Adolf Eichmann, where many survivors of the Holocaust testified. For the first time, Israelis publicly heard first-hand accounts of the Nazi atrocities given by the victims themselves. The second event was the trauma of the Yom Kippur War of 1973, which made the Israelis feel—more than ever before—vulnerable and exposed. To these factors we should add others:

> Such developments as the decline of Labour Zionism and the rise of the Likud government, the growing political and cultural impact of the more traditionally oriented Israelis of Middle Eastern descent, the greater role of religion in Israel national culture, and the closer contact between Israelis and Jewish communities abroad began to transform Israeli political culture and shake earlier views expressed in the binary oppositions of Exile/the Land of Israel, Hebrews/Jews, secular nationalism/religious tradition. [Zerubavel, 1994, p. 86]

The Holocaust was subsequently reconsidered within the context of Israel's political present: considerable empathy was felt towards the victims of Nazism, and a closer process of identification with them unfolded. Consequently, the historical narrative describing Masada changed: instead of highlighting the contrast between Masada and the Holocaust, at this point "the importance of the *suicide as the tragic climax* of an extreme state of besiegement and persecution" (*ibid.*, p. 87) is given special attention in the literature. If suicide became an unavoidable choice when Jews were sadistically persecuted and cruelly victimized, then this kind of national tragedy should clearly be avoided. Both Masada and the Holocaust should never be repeated. The "fragility of Jewish survival" was thus stressed, legitimizing the current political concern for Israel's security. Today, the soldiers of the armoured units of the Israeli Defence Forces swear the oath of allegiance on the summit of the fortress.

Let us pause at this point. Memory, though it contains fracture and loss, defines the relationship between past, present, and future. *Les Lieux de Memoire*, as Pierre Nora (1989) has called them, those "places" where memory and history interact in reciprocal overdetermination, will reflect "past" history as much as the "present". Be it Proust's *petite madeleine* or a monument like Masada, an individual or a group can and (consciously and / or unconsciously) will use the representation of their past according to the needs of the present. Individual and / or group narratives that commemorate the past will change over time, according to circumstances. Every memory is an interpretation, a new reconstruction of the past. Contemporary biologists and neurologists now believe that every time an object of experience is recalled in our minds, memories emerge from an impossible maze of neural firing formations and synaptic connections. The process of remembering is changed every time by a complex interaction of new association and experiences. Politics greatly depends upon such readings and revisions. Foucault (1975) has argued that "If one controls people's memory, one controls their dynamism . . . It is vital to have possession of this memory, to control it, administer it, tell it what it must contain" (pp. 25–26, quoted in Baker, 1985). As we have just seen, the symbol of Masada changed as a piece of commemorative narrative in the history of the Jewish people.

Freud always believed that every subject, as well as every nation, revises past events at a later date; this revision is what creates a historical past, what gives those events meaning. The first reference to this notion in Freud's writings appeared in the article on *screen memories*, where he described two different kind of movements: in the first case, an early memory is used as a screen for a later event; the screen memory here is "pushed forward". In the second case, a later memory is used to screen an early event; it is "retrogressive" (Freud, 1899a). While the first type will later disappear from Freud's writings, the second one will be called *retrospective phantasy* in the *Interpretation of Dreams* (1900a). The concept will thereafter reappear in *The Psychopathology of Everyday Life* (1901b), and play an important part in Freud's arguments in connection with the concept of the primal scenes in the case of the Rat Man (1909d); the Wolf Man (1918b); and in the *Introductory Lectures* (1916–1917).

Nevertheless, Freud's concepts should be clearly differentiated from Jung's notion of the neurotic revision of the past: the adult, according to Jung, reinterprets his past in his phantasies of reconstruction, which are, in fact, a symbolic expression of his present troubles. At an individual level, it is a kind of defensive manoeuvring, the expression of a wish to escape from neurotic difficulties of the present to an imaginary past. While for Jung the neurotic invents a past in order to escape his present, for Freud this is an incomplete version of what actually happens. The neurotic invents a past for himself, but it is *what has happened in the past, whether phantasized or in reality, that explains such an invention.* Here resides the relevance of the Freudian concept of *nachträglich,* which destroys any hope of believing that history follows a linear determinism, i.e., that the present is caused by the past.[6]

Let us at this point go back to Masada. Let us consider the impact of Masada as a representation of Jewish identity, and its possible influence on the political views of contemporary Israelis. Even taking into account that Josephus was not a reliable historical source, it is nevertheless significant that the events at Masada have been *imagined,* that is, that they have been conceived as *possible.* It is not uncommon today to find references to the "Masada mentality" in the popular press in Israel. One finds in the professional literature authors who argue for the existence of a "Masada complex" in the mentality of the people of Israel.[7]

Are the Israelis, their government, some of their leaders and politicians, unconsciously identifying a bit too much with those that defended Masada, with the Zealots who killed their families and then committed suicide? The fanatical Orthodox settlers are comparable to the Zealots. Like the ancient sect, they, too, refuse to follow the rules established by law; instead, their ideas are dictated by their theological and messianic beliefs. In contrast to the predominance of learning and the passion for justice present in the Talmud, they have ruthlessly subscribed to a number of self-intoxicating heroic myths. While they have not actually started a war (as the Zealots did against the Romans through their chauvinistic arrogance and an ideology of racial exclusivity), they have nevertheless actively participated in provoking and prolonging the cycle of extreme violence. The fanatics of today may not have massacred their Jewish opponents but they have created an atmosphere of civil war within Israel. The Zealots practised certain forms of terrorism; their modern counterparts have inspired the murder of one Israeli Prime Minister and many acts of unbridled aggression against Palestinians.

Is an unconscious identification (with the Zealots of the past and seemingly with the settlers of today) compelling the Jews of Israel to repeat the plot of a mythical story? Is a compulsion to repeat blinding them and preventing them from finding a peaceful solution to their conflict with the Arabs? The birth and the very existence of Israel as a nation have been represented within a similar context: David against Goliath, a tiny group of Jews against the might of the Arab nations. This mythological construction declares that if Israel has come to exist *because* of Masada, the Jews of Israel continue to survive *because* of those heroic past events. Above all, and more worryingly, they will *only* survive in the future if they go on "fighting to the end", instead of negotiating a fair peace with the Palestinians. It is striking, and rather uncanny, to notice how much all Israel's political parties, even considering their internal divisions, feed and survive on this siege mentality. Many of Israel's political leaders, both from the left and the right, represent Israel in their political statements as "Masada". They continue to picture themselves as a group of vulnerable Jews, surrounded by enemies, isolated on top of a mountain.

IV

There is no acceptable way to simplify or easily generalize about historical events. Historical processes do not conform to given laws that are understood or proven, and the arguments presented in this chapter can only add to the complexity of the "dark mass of factors" (Berlin, 1996, p. 37) that determine historical events. At present, historians agree that there is one unfortunate lesson to be learnt from historical experience: that nobody ever seems to learn from it. Psychoanalysts, from their own experience gained in their consulting-rooms, would have plenty of reasons to agree with this too. And yet, this does not imply that we should give up hope. On the contrary, we should go on opposing ignorance and political repression, continue fighting against individual and social cruelty, denouncing injustice and superstition. This is not a task to be left to politicians. *Certainly not* to politicians alone.

My point, in the end, is a limited one: things that happened in the (mythical or historical) past, which were not possible to incorporate in a meaningful context at the time (thus, they were traumatic), will be revised *nachträglich* so as to give significance to them *a posteriori* in the present. This creation of a meaningful past (what Freud (1887–1904) called "a retranscription") contributes to structure the invention of the present; that is, it determines the way we perceive the world and how we know it, the way we construct our knowledge of the present. This in turn (and sometimes dangerously) might determine the construction of the future through the compulsion to repeat.

Notes

1. For a discussion of the subtlety and complexity of Freud's concept see—among others—Laplanche and Pontalis's comments (1967); the article by Helmut Thomä and Neil Cheshire (1991); the note included at the end of Laplanche's collection of papers in English (Fletcher & Stanton, 1992); and Modell's book (1990). I have briefly expressed my criticisms of Breen's understanding of this concept in the Bulletin of the Psycho-Analytical Society (Kohon, 2003).
2. Similarly, the traditional people of New Zealand, the Maori, thought of the future as *kei muri*, "behind" them, while the past was *nga ra o mua*, "the days in front". For them, past and future were always *present*. The Maori's conception of time also played an important role in the encounter between the Maori and their conquerors, and the mutual misunderstandings in the reading of the signs between these two groups (Sahlins, 1985)
3. For the events at Masada, and the comments included in the following pages, I consulted—among others—the following authors: Dwyer (1979); Funk (1974); Josephus (1970); Schwartz and Kaplan (1992); Spero (1978); and especially Yadin (1966) and Zerubavel (1994).
4. Inscribed or painted fragments of pottery or limestone flake used in antiquity, especially by the ancient Egyptians, Greeks, and Hebrews.
5. Halbwachs also argues that history is always written, while memory is not; that as long as the past survives in the collective memory, there is no difference between past and present, and that history is not necessary. In his view, history would come into existence only at the point of social memory fading.
6. Nelson Mandela begins his autobiography thus:

   > Apart from life, a strong constitution and an abiding connection to the Thembu royal house, the only thing my father bestowed

upon me at birth was a name, Rolihlahla. In Xhosa, Rolihlahla literally means "pulling the branch of a tree", but its colloquial meaning more accurately would be "troublemaker". I do not believe that names are destiny or that my father somehow divined my future, but in later years, friends and relatives would ascribe to my birth name the many storms I have both caused and weathered. [Mandela, 1996, p. 3]

7. There have been many versions of the concept of complex, but in *Freudian* psychoanalysis, complex is understood in a restricted sense, as a basic unconscious structure that, based on the history of an individual, results in the constitution of an organized group of ideas that determines the way an individual chooses his present, his future, his neurosis, or his self (cf. Laplanche & Pontalis, 1967).

References

Baker, K. M. (1985). Memory and practice: politics and the representation of the past in eighteenth-century France. *Representations, 11* (Summer): 134–164.

Berlin, I. (1996). *The Sense of Reality—Studies in Ideas and their History*. London: Pimlico.

Berlin, I. (1997). *The Proper Study of Mankind—An Anthology of Essays*. London: Pimlico, 1998.

Borch-Jacobsen, M. (1996). *Remembering Anna O.—A Century of Mystification*. New York & London: Routledge.

De Certeau, M. (1986). *Heterologies: Discourse on the Other*. Manchester: Manchester University Press.

Clendinnen, I. (1991). "Fierce and unnatural cruelty": Cortés and the conquest of Mexico. *Representations, 33* (Winter): 65–100.

Collingwood, R. G. (1946). *The Idea of History*. London: Oxford University Press.

Darnton, R. (1984). *The Great Cat Massacre—And Other Episodes in French Cultural History*. Harmondsworth: Penguin, 1985.

Dilthey, W. (1894). *Ideen über eine beschrcibende und zergliedernde Pyschologie*. Gesammelte Schriften 5. Leipzig: Teubner, 1924.

Dwyer, P. M. (1979). An inquiry into the psychological dimensions of cult suicide. *Suicide and Life-Threatening Behavior, 9*(2): 120–127.

Fletcher, J., & Stanton, M. (Eds.) (1992). *Jean Laplanche: Seduction, Translation and the Drives, A Dossier edited by John Fletcher and Martin Stanton*. London: ICA.

Foucault, M. (1975). Film and popular memory: an interview with Michel Foucault. *Radical Philosophy, 11*: (Summer).

Freud, S. (1887–1904). *The Complete Letters of Sigmund Freud to Wilhelm Fliess—1887–1904*. J. M. Masson (Ed. and Trans.). Cambridge, MA: Harvard University Press, 1985.

Freud, S. (1899a). Screen memories. *S.E., 3*. London: Hogarth.

Freud, S. (1900a). *The Interpretation of Dreams. S.E., 4–5*. London: Hogarth.

Freud, S. (1901b). *The Psychopathology of Everyday Life. S.E., 6*. London: Hogarth.

Freud, S. (1909d). *Notes Upon a Case of Obsessional Neurosis. S.E., 10*. London: Hogarth.

Freud, S. (1916–1917). *Introductory Lectures on Psycho-Analysis. S.E., 15–16*. London: Hogarth.

Freud, S. (1918b). *From the History of an Infantile Neurosis. S.E., 17*. London: Hogarth.

Funk, A. A. (1974). A Durkheimian analysis of the event at Masada. *Speech Monographs, 41*: 339–347.

Gay, P. (1974). *Style in History*. New York: Norton, 1988.

Gay, P. (1978). *Freud, Jews and other Germans—Masters and Victims in Modernist Culture*. New York: Oxford University Press.

Green, A. (1993). *Le travail du négatif*. Paris: Les Editions de Minuit.

Halbwachs, M. (1980). *The Collective Memory*. New York: Harper & Row.

Hobsbawm, E. (1994). *Age of Extremes—The Short Twentieth Century 1914–1991*. London: Abacus, 1995.

Hobsbawm, E. (1997). *On History*. Weidenfeld & Nicolson, London.

Johnson, P. (1987). *A History of the Jews*. London: Weidenfeld and Nicolson.

Josephus, F. (1970). *The Jewish War*. G. A. Williamson (Trans.). Harmondsworth: Penguin.

Klauber, J. (1981). *Difficulties in the Psychoanalytic Encounter*. New York: Jason Aronson.

Kohon, G. (1999). *No Lost Certainties to be Recovered*. London: Karnac.

Kohon, G. (2003). Brief comments on Dana Birksted-Breen's paper "Time and the *Après Coup*". *The Bulletin of the British Psychoanalytical Society, 39*(1): 17–18.

Kohon, G. (2005). Love in a time of madness. In: A. Green & G. Kohon (Eds.), *Love and its Vicissitudes*. London: Routledge.

Laplanche, J., & Pontalis, J.-B. (1967). *The Language of Psychoanalysis*. London: Hogarth.

Mandela, N. (1996). *Long Walk to Freedom—The Autobiography of Nelson Mandela*. London: Abacus.

Modell, A. (1990). *Other Times, Other Realities—Toward a Theory of Psycoanalytic Treatment*. Cambridge, MA: Harvard University Press.

Nora, P. (1989). Between memory and history: *Les Lieux de Mémoire*. *Representations*, 26:Spring 1989: 7–25.

Sahlins, M. (1985). *Islands of History*. Chicago, IL: University of Chicago Press.

Schwartz, M., & Kaplan, K. J. (1992). Judaism, Masada, and suicide: a critical analysis. *Omega, 25*(2): 127–132.

Segal, H. (1991). *Dream, Phantasy and Art*. London: Routledge.

Spero, M. H. (1978). Samson and Masada: altruistic suicides reconsidered. *The Psychoanalytic Review, 65*(4): 631–639.

Thomä, H. (1989). Translation in transition: the case of Sigmund Freud and James and Alix Strachey. Paper given at The Institute of Psychoanalysis Conference, London, 20–22 April.

Todorov, T. (1982). *The Conquest of America—The Question of the Other*. New York: Harper & Row, 1984.

Yadin, Y. (1966). *Masada—Herod's Fortress and the Zealots' Last Stand*. London: Weidenfeld and Nicolson.

Yerushalmi, Y. H. (1982). *Zakhor, Jewish History and Jewish Memory*. New York: Schocken Books, 1989.

Zerubavel, Y. (1994). The death of memory and the memory of death: Masada and the Holocaust of historical metaphors. *Representations, 45*: (Winter): 72–100.

CHAPTER SEVEN

Borges, immortality, and "The circular ruins"[1]

Catalina Bronstein

Borges

Jorge Luis Borges was born just over a century ago, on 24 August 1899 in Buenos Aires. He died in Geneva on 14 August 1986. The Borgeses were sons and grandsons of officers who served in the War of Independence against Spain, in the campaigns against the Indians, in nineteenth-century wars against Argentina's neighbours, as well as in Argentinean civil wars. Borges's grandfather was a colonel who is said to have died heroically in one of the civil wars. His paternal grandmother was English, and Borges was raised bilingually. This combined upbringing, where English authors mingled with Spanish culture, as well as the cultural changes following the massive immigration of Italians at the beginning of the century, together with the historical influence of the building of Argentina, produced the rich blend of experiences from which Borges would draw.

When Borges was fourteen, his family moved to Geneva searching for treatment for his father's eye problems; these were hereditary and had left several members of his family blind. His father died blind, as did Borges. Borges finished his secondary education in Switzerland and spent time in Spain before returning to Buenos Aires (Bell-Villada, 1981). Borges regarded himself primarily as a poet until 1938, when two very painful episodes in his life seemed to come together to produce a vital change in his writings: first, his father died and, some months later, Borges suffered an accident on a dark stairway. He had gone to fetch a young woman whom he was going to take out and he did not see an open window. His serious head-injury developed into a septicaemia that kept him between life and death for fifteen days. It was after recovering from this delirious ordeal that Borges wrote what became his first work of metaphysical fiction, *Pierre Menard, Author of the Quixote* (Barnatan, 1995; Bell-Villada, 1981; Salas, 1994; Woscoboinik, 1991). It is not difficult to speculate on the importance that these very painful events may have had on his writing, though it is not the aim of this paper to undertake a psychoanalytic study of Borges as a person or to search for possible unconscious motives for his ideas.

Borges's stories are surrounded by a magical aura that seems to be created by the "fiction-within-a-fiction". In a study on Borges, Bell-Villada suggests that this "derives from the possibility that we, the readers of outer fiction, may be in turn fictional characters being read about, at the very same time, by someone else!" (Bell-Villada, 1981). Borges plays with fact and fiction, the real and unreal, illusion, time, and death. This is appropriately described by one of the characters in his story 'Tlon, Uqbar, Orbis Tertius', who states that "metaphysics is a branch of fantastic literature" (1944a).

"The circular ruins"

"The circular ruins" (1944b) is a short, strangely elusive and highly poetic story about the struggle of an individual's determination to generate a man. In this story, an old man, a magician, gets off a boat, kisses the "sacred mud" and installs himself in the ruins of an ancient circular temple. He is moved, compelled by the "invincible", "not impossible", but supernatural purpose "to dream a man: he wanted to dream him with minute integrity and insert him into reality". We are told that this man probably does not feel any physical pain: "This magical project had exhausted the entire content of his soul; if someone had asked him his own name or any trait of his previous life, he would not have been able to answer".

The Old Man dedicated himself to dreaming and he dreamed "that he was in the centre of a circular amphitheatre which in some way was the burned temple . . ." He chose a boy from among others in his dreams. This boy had "sharp features which reproduced those of the dreamer". Borges continues, stating that "nevertheless, catastrophe ensued", as the man could no longer go to sleep. Eventually, and after worshipping the planetary gods and uttering the syllables of a "powerful name", he slept. He finally dreamed a complete man; one afternoon, he almost destroyed his work, but then repented. Borges writes, "It would have been better for him had he destroyed it". Finally, the Old Man makes a pact with the God of Fire (who presides over the temple), who "would magically give life to the sleeping phantom, in such a way that all creatures except fire itself and the dreamer would believe him to be a man of flesh and blood". Each day, the Old Man prolonged the hours he dedicated to his dreams as it "pained him to be separated from his boy". But the boy was now ready to be sent to the other circular temple. "But first (so that he would never know he was a phantom, so that he would be thought a man like others) he instilled into him a *complete oblivion* of his years of apprenticeship" (my italics).

Borges tells us that this man went on living, imagining that his unreal child was practising the same rites as him in other circular ruins. "His life purpose was complete; the man persisted in a kind of ecstasy." Some undetermined time later, the dreamer was awakened by two men who told him of a magic man who could walk on fire and not be burned. This tormented the Old Man, who feared his son would discover "that his condition was that of a mere image . . . Not to be a man, to be the projection of another man's dream, what a feeling of humiliation, of vertigo!"

Suddenly, fire begun to ravage the old man's own temple. And Borges writes,

> For what was happening had happened many centuries ago. The ruins of the fire god's sanctuary were destroyed by fire. In a birdless dawn the magician saw the concentric blaze close round the walls. For a moment, he thought of taking refuge in the river, but then he knew that death was coming to crown his old age and absolve him of his labours. He walked into the shreds of flame. But

> they did not bite into his flesh, they caressed him and engulfed him without heat or combustion. With *relief*, with *humiliation*, with *terror*, he understood that he too was a mere appearance, dreamt by another [p. 76, my italics).

The illusion we are under in this story is the idea that the Old Man makes a voluntary choice and that he can actually make a decision—to destroy his work, to stop the dreaming. What Borges shows us is this Old Man locked, encircled by, and inside the ruins of his own making. How can he step outside his dream of dreaming a man, if he himself is the product of somebody else's dream? Is it just an act of volition to accept reality or is it that, at times, somebody might be unable even to see that he is the product of a dream; not just somebody else's dreams but also his own dream about how he was dreamed or wanted to be dreamed by somebody else? And couldn't this be a way of avoiding the painful knowledge of his mortality?

The story's epigraph is taken from *Through the Looking Glass* by Lewis Carroll: 'And if he left off dreaming about you . . .' It is from a scene in which Tweedledum and Tweedledee tell Alice about the Red King. They explain that the King is actually dreaming of her and, were he to awaken, Alice would simply vanish. Tweedledee asks Alice:

> 'And if he left off dreaming about you, where do you suppose you'd be?'
>
> 'Where I am now, of course,' said Alice.
>
> 'Not you!' Tweedledee retorted contemptuously. 'You'd be nowhere. Why, you are only a sort of thing in his dream!' [Carroll, 1911, p. 165]

Alice is thus denied a real existence, a real identity, and becomes someone else's dream. Maybe Alice herself was creating a dream of being somebody else's dream. In Borges's story, were the dreamer to awaken, his son would vanish, but it is quite likely he would then also vanish, being himself just a dream and therefore also lacking a real identity.

With the help of the gods he invokes, the Old Man reproduces himself and creates a man in his own image. The son should be the same as the father and is created with the help of another father,

God. It is a major narcissistic endeavour that defies gender differences, time, history, and human change.

This same subject is brought up again in another beautiful and quite complicated story called 'The immortal' (1949), from Borges' book *The Aleph*.

"The immortal"

The story opens with a note on Joseph Cartaphilus, an aged antique dealer who recently sold a princess an original edition of Pope's translation of the *Iliad*. Inside the book she discovers a manuscript, which is the story one then reads. In this story, an army officer called Marcus Flaminius Rufus, who had served in a legion where the men were consumed by "fever and magic", precipitously decided to seek out the famous City of Immortals and the River of Immortality, "the secret river which cleanses men of death". The reasons behind his action are vaguely described as a possible sense of privation for being unable to have a proper look at the face of Mars. After enduring much hardship, Rufus found himself lying with his hands tied, close to the River of Immortality and to the City of Immortals. He was amid "naked, grey-skinned, scraggly bearded men who could not speak and who ate serpents: the troglodytes" (cave dwellers). He finally reaches the City, followed by one of these troglodytes, who remains outside.

Rufus has difficulties in finding the way to the centre of this City. He has to work his way out of a vast circular chamber with nine doors, eight of which lead to labyrinths that return to the same chamber. He finally emerges into the courtyard of an empty, "interminable", "atrocious", and "completely senseless" building, older than mankind. The architecture lacks any finality: "It abounded in dead-end corridors, high unattainable windows, portentous doors which led to a cell or pit, incredible inverted stairways whose steps and balustrades hung downwards".

Rufus thinks to himself: this palace is a fabrication of the gods. Then he corrects himself: the Gods who built it have died. He then continues to say: the Gods who built it were mad.

Rufus leaves the place and tries to teach language to the troglodyte who waited for him, and whom he had called "Argos" because of his hang-dog quality. He fails over and over again. Borges writes,

> I thought that Argos and I participated in different universes . . . I thought that perhaps there were no objects for him, only a vertiginous and continuous play of extremely brief impressions. I thought of a world without memory, without time . . . Thus the days went on dying and with them the years, but something akin to happiness happened one morning. It rained, with powerful deliberation. [p. 143, my italics]

Rufus dreams that a river (to whose waters he has returned a goldfish) has come to rescue him.

> Argos, his eyes turned towards the sky, groaned; torrents ran down his face, not only of water but also (I later learned) of tears. Argos, I cried, Argos. [*ibid.*]

Suddenly Argos stammers a line out of the Odyssey, "Argos, Ulysses's dog". When Rufus asks him how well he knows the *Odyssey*, the once mute troglodyte says, "It must be a thousand years since I invented it". He was Homer himself. The mad city was the "last symbol to which the Immortals condescended; it marks a stage at which, judging that all undertakings are in vain, they determined to live in thought, in pure speculation . . . Absorbed in thought, they hardly perceived the physical world". *Except for man, all creatures are immortal,* for *they are ignorant of death; what is divine, terrible, incomprehensible, is to know that one is immortal* (p. 144, my italics).

Borges also tells us in the words of Rufus:

> Homer composed the *Odyssey*; if we postulate an infinite period of time, with infinite circumstances and changes, the impossible thing is not to compose the *Odyssey*, at least once. No one is anyone, one single immortal man is all men. Like Cornelius Agrippa, I am god, I am hero, I am philosopher, I am demon and I am world, which is a tedious way of saying that I do not exist. [p. 145]

Immortality "*made them invulnerable to pity . . . neither were they interested in their own fate*" (my italics). "Among the immortals . . . every act (and every thought) is the echo of others that preceded it in the past . . ."; "Death (or its allusion) makes men precious and pathetic" (p. 146).

Rufus decides to search for the river whose waters remove immortality. After many more years Rufus sees a spring of clear water, which he tastes. Some time later, a spiny bush lacerates the back of his hand. The unusual pain seems very acute and makes Rufus say, "*Incredulous, speechless* and *happy*, I contemplated the precious formation of a slow drop of blood. Once again I am like all men" (p. 142, my italics).

But the end brings in an ironic negation of this same statement. Mortality seems to be best, but Borges goes back to stress how, when the end draws near, there are no longer "any remembered images; only words remain".

Borges presents us with two very imaginative stories, the product of a highly creative enterprise. Among many other possible interpretations, we can understand his narrative as a metaphor for the psychic struggle experienced by some individuals when they are confronted with feelings of pain and loss and their subsequent compulsive search for an omnipotent solution through the phantasy of becoming immortal. In a masterly way, Borges describes the mixed feelings experienced by the two men in the stories when they become aware of the omnipotent but deadly consequences of becoming immortal: "Relief, humiliation and terror" in "The circular ruins", and a "sense of the divine, incomprehension and madness", as well as the loss of language as meaning, as communication, and as vehicle for individuation, in "The immortal".

Both stories deal with the omnipotent phantasy of cancelling chronological, linear time. Differences are cancelled; there is no identity to separate the individual from anybody else, no different identity to make the son different from the father. Procreation can happen without sexual intercourse between a man and a woman. There is no physical body to get anxious about and attend to. Basically, no other objects are allowed to exist in the mind, just echoes of oneself. In "The immortal" Borges brilliantly describes the result when he says, "There is nothing that is not as if lost in a maze of indefatigable mirrors".

Anzieu stresses that the child enters into chronological time only through the acceptance of the "forbidden oedipal other". For him, the "circular time" in Borges is a symbolic figure of the circle inside which mother and child fuse. It is a time of the indefinite repetition of pleasure (Anzieu, 1971). I agree with Anzieu in that the circularity symbolizes a closed system where there is no room for any separate objects. There is certainly no room for a primal scene. These are ruins that I think represent the Old Man's compulsion to repeat "this magical project [that] had exhausted the entire content of his soul" (Borges, 1944b). However, neither the Old Man nor Rufus seem to be seeking pleasure alone. They seem possessed by a need, which they do not feel they have much choice over.

Mr A

I would like now to describe a patient, Mr A, who, like the Old Man, is locked in a circular temple and for whom coming to analysis seems to be equivalent to Rufus's conflict about recovering his own mortality.

> We can think of Mr A, a man in his thirties, as somebody who, like Rufus, is searching for a miraculous river that would cancel reality. Mr A had a very difficult childhood with both parents chronic alcoholics who drank themselves into comas and finally died debilitated by drink and liver failure. However, Mr A was left with the image of a grandiose father whom all his friends admired and who came alive through drinking. His father used to talk about his numerous sexual conquests, even though the boy had to hear him confessing later that he was impotent. His mother was felt by him to be somebody who renounced everything and drank herself to death in her need to follow an idealized and feared husband.
>
> As a child, Mr A got involved in dangerous playing which prompted his numerous accidents, with several hospitalizations for broken bones. He was not aware that he might have wanted to die or that he could have died. He walked into a lake without knowing how to swim and had to be rescued. He fell from swings, trying to get higher and higher, walked on the edge of dangerously high walls, and so on. As an adult he gambles with alcohol and feels terribly surprised when he urinates in his sleep, and when, the day after drinking heavily, his mind and body do not feel quite right. He cannot believe that they have suffered from the impact of drinking.
>
> After his father's death, which occurred during the first year of his analysis, Mr A turned to drinking more heavily than before. He

frequently fell asleep in the sessions and he missed a number of these after having gone drinking the night before. His associations invited interpretations about loss, but I was often left feeling that I had been seduced into talking about pain and loss when he could not feel either of them. At this time I felt that there was a demand coming from him that I should carry the anxiety and pain about the loss of his father, and about his own self-destructive behaviour, while he observed my impotent struggle to help him get in touch with an emotional experience that he had magically made disappear while drifting into a world without time, without memory. He often spent his sessions organizing geometrical pictures in his mind, controlling, and mostly cancelling, any impact that my words provoked in him.

Three months after his father's death, he said, "Yesterday was a strange day. I just wanted to sleep. I remember being at the tube station wishing that I could go and get into bed, but I went to work instead. I went back home early and got into bed. I set the alarm at six and I was slightly panicky when it went off because I did not know whether it was six in the evening or six in the morning but then I went back to bed and went on dozing. I cannot remember what we talked about yesterday. I remember I said something about trousers and about my boss but cannot remember what you said, although I felt surprised by it. I think I was expecting you to say something and you said something else."

This was said in an affect-less way and it seemed like a precursor to his going to sleep. I took up his wanting me to keep the session alive and remember for him what he had forgotten. My bringing up his father's death yesterday alarmed him, so he cancelled the memory of the session by going to sleep, as he was about to do now as well.

"It is strange you say that because I have completely forgotten that my father died. I now remember thinking a bit about it while I was outside . . . I am aware when I meet people that they are aware that he died and that they say things like how much they sympathize with me. When he died I was with my brother and sister and they both had red eyes and they both said that they had cried their eyes out and then felt better and I felt robbed of my tears. Yesterday, before going to sleep, I decided to masturbate and looked for a fantasy but nothing turned me on. I thought, Am I turning asexual?"

He yawned. He said he has been struggling, trying not to fall asleep. After a silence he said, "I have been looking at that picture on the wall. In the picture it is dawn. Maybe that has to do with goodbyes. That word strikes a chord. I have been playing one of my games just now, imagining that all the angles in the picture were changed."

His "game" seems to be about changing the chords of an emotional experience so as to produce meaningless pictures, which he then watches from a distance, only to feel that he has been robbed of his tears. Thus, awareness of anxiety at his attack on his capacity to perceive time (which would have inevitably led to an experience of loss and pain) was cancelled out.

In his second year of analysis I learned that Mr A leads a secret mental life of day-dreams. In these day-dreams he is a hero saving people from disasters. His friends, his analysis, his work, all seem to be just minor necessary undertakings which never feel as gratifying as his day-dreams. In these day-dreams Mr A gives birth, like the Old Man, to a him who is invulnerable; to a him who is immortal as well as superior to all other human beings. This him does not suffer pain or fear, is ready to risk giving up his life to rescue others, but without ever losing it. He is a real Superman, morally and physical superior to all of us. Being deeply religious, Mr A feels at times at one with God and he has intense sensuous experiences in which he feels a "call" from God.

As soon as he leaves the session, even before the door closes, Mr A is off saving a child from some possible terrible disaster. He is not consciously aware that he is the child who needs to be saved, as he has the proof that he can be harmed, hurt, almost dead, and yet he never dies. He feels powerful and, in the brief moments he looks around and sees other people going to work, he feels sorry for them. Mr A lives somehow like the troglodyte. There are no real objects that seem to matter. He could not really mourn his parents' death; he cannot enjoy human intercourse. He does not know whether he is homosexual or heterosexual. Ordinary women are felt to be interesting and desirable as long as they are unreachable, and he has a number of idealized women whom he keeps at a distance, away from any possible development of a real relationship (this being also enacted in the analysis).

When Mr A gets in touch with his reality he feels terrified that he will be "found out" and, in his mind, the ordinary human him would then be despised by others in the same way as it is despised by him. In fact, Mr A is terrified that he will "find out" about his attacks on the real human him and of his compelling need to repeat a story in his mind that leaves him feeling empty, lonely, and despairing and that could eventually lead to his actual death. At the same time, in his coming to analysis, he seems to be searching for a safe place where he could allow himself to be "found out". Mr A sought analysis out of anxiety that he was going mad. He was in five-times-a-week analysis for ten years.

Though from his point of view Mr A feels no desires but is aware of others who seem to need things from him, it is clear that at a deeper level he projects his own desires and needs into other people, who then seem to represent to him these aspects of himself. Therefore, like the Old Man, he does not feel he is living his own life. He is always the product of his own dreams and of what he phantasizes were his father's dreams. He is seldom aware of any feelings, or that he is the one doing the dreaming. During the first years of analysis the transference was imbued with a demand for me to follow him and get lost with him in the labyrinths he created. At the same time I was exposed to the anxiety that he could not feel about his self-destructive behaviour. I was placed in the position of a spectator, probably similar to how he must have felt as a child, an impotent witness to his parents' suicidal behaviour. It seemed vital that I should not become part of his fiction and that the analysis could help him discover "death within himself, not death wishes or anxiety, but 'deathwork'", as it has been described by Pontalis (1977), and offer him an alternative to his belief that only his omnipotence could save him.

In the fifth year of his analysis, Mr A brought the following dream to one of his sessions: "I dreamed I was on a boat. The captain ordered one of the men on board to do something and then told him he would not be able to be part of the crew any longer. I thought this was unfair. He was talking to me and I could see his face, like that of a drunken man with red blood vessels. Then I was in a large vessel, a liner, and I was under the deck and the boat first rolled on to one side and then turned round completely. It did the same to the other side. I felt everything was going round and round. I realized we were sinking. I thought I should get a bucket to put over my head so I could breathe when I got to the bottom of the sea and I would then survive. I looked for a bucket and there was a pile of buckets, champagne buckets that you see in pubs. I put one over my head and went down.

"I could therefore breathe so I stayed there and then I came up. I saw people trying to get hold of things and there was a couple that had a baby who had died. Then I thought of Louise and I remembered that I was with her and I started looking for her. I started crying and crying, until I woke up."

I will return to this dream later.

Death and delusion

Borges's description of both the Old Man's and Rufus's world is one that reminds us of Freud's description of the workings of the primary process in the System Unconscious: mainly wishful impulses, "impulses existing side by side exempt from mutual contradictions", "no negation, no doubt, no degrees of uncertainty", no reference to time and therefore timelessness, no regard for reality (Freud, 1915e).

In "Thoughts for the times on war and death", Freud states that it is impossible to imagine our own death and that "at bottom no one believes in his own death . . . and that in the unconscious every one of us is convinced of his own immortality" (1915b). But he says,

> Man could no longer keep death at a distance, for he had tested it in his pain about the dead; but he was nevertheless unwilling to acknowledge it, for he could not conceive of himself as dead. So he devised a compromise: he conceded the fact of his own death as well as *denied it the significance of annihilation*. [1915b, p. 299, my italics]

He adds that man needed religions to present a possibility of an after-life in order to deprive "death of its meaning as the termination of life". We can speculate that in both stories, as well as in Mr A, the fear of death might have been the trigger for the denial of chronological time and for the search for immortality. However, I think that this is an insufficient explanation.

In 1914, Freud developed the concept of narcissism as a withdrawal of libido from the external world and directed to the ego (Freud, 1914c). Rosenfeld (1971) explored this subject further, focusing on the central role played by the over-valuation of the self, based mainly on the idealization of the self. As he sees it, there are omnipotent parts of the self, which may be split off from the rest of the personality and become like a "delusional world or object into which parts of the self tend to withdraw". He stresses that this omnipotent and omniscient part of the self "creates the notion that within the delusional object there is complete painlessness" and sees it as a "power which manages to pull the whole of the self away from life into a deathlike condition by false promises of a Nirvana-like state" (*ibid.*, p. 175). He adds that in this process the saner self enters the delusional object through projective identification and loses its identity, becoming completely dominated by this omnipotent destructive process.

This delusional object that carries the omnipotent aspects of the self seems to fulfil a parental role on which the subject later feels dependent in order to survive, like the Old Man in "The circular ruins", who turned to "the planetary gods, uttered the lawful syllables of a powerful name and slept" (Borges, 1944b). The real parents are denied existence in the same way as the real and vulnerable aspects of the self; they have been killed and replaced by a delusional system. What is left of them is just circular ruins. The Old Man in "The circular ruins" disembarked in what was the remains of a temple "long ago devoured by fire".

Attacks on the awareness of a human non-ideal mother, whom the infant does not own, can lead, according to Brenman, to an identification with the "ideal breast" that would satisfy the demands to have the ideal and be the ideal. The outcome can be the development of a superego that is cruel and narrow and that the individual feels forced to worship and satisfy for the rest of his life. The individual would therefore be "confined to his narrow loveless narcissistic demands, governed by narrow loveless narcissistic gods" that are loved more than humanity (Brenman, 1985).

The triumph experienced towards the vulnerable and fallible human parental objects as well as towards the vulnerable aspects of the self, and the rejection and attack on both of them, brings about an inner sense of persecution and fear of retaliation. This fear of retaliation is often at the basis of the fear of death, as was stated by Freud: "The fear of death . . . is usually the outcome of a sense of guilt" (1915b, p. 297). Thus, the delusional narcissistic withdrawal that has led to triumph over the parts of the self aware of pain and vulnerability generates persecutory guilt and fear of death, which needs to be dealt with by further withdrawal into this omnipotent state of mind. These are the circular ruins that cannot be abandoned. The "painless world" that has been created has to be maintained at all costs as the experience of psychic pain is equated to death.

In "The circular ruins" the Old Man created the gods he needed to invoke in order to pursue his quest for immortality. He was also dependent on the gods' will. These gods protected his narcissism but also enslaved him to it. He had to submit to the fate of seeing the destruction of the temple but could not completely abandon it, as he was trapped inside its ruins and could not die. He was left without any real resources and, as Borges says (in relation to the Old Man), his project had "exhausted the entire content of his soul" and "if someone had asked him his own name or any trait of his previous life, he would not have been able to answer" (Borges, 1944b). The individual is left impoverished, with no external or internal separate objects to relate to except his godlike omnipotent double.

The actual ruins would symbolize in this way the remains of his dead objects as well as his own identity and his feeling of being trapped by the repetitive compulsion to go on recreating the same scenario. This reminds us of Freud's development of his theory of the death drive (1920g) as opposed to a life drive. A death drive that would explain situations where the individual tends to repeat unpleasant and traumatic experiences, which might induce suffering but which he cannot avoid repeating; a death drive that "pushed the organism to die only in its own fashion" and that is linked by him to destructiveness and repetition compulsion, thus entering into a conflict with the wish to live, to love, and to create. Zilboorg ventured the possibility that "the death instinct projected outward might not only become the instinct of eternal aggression but also the drive towards immortality" (1938). In the case of the Old Man, as well as in Mr A, it seems to lead to the cancellation of psychic pain inherent in the knowledge of time, change, and differences; more akin to how Aulagnier describes her conception of the death drive—the desire for not having to entertain any desire (1975). In Mr A this amounts to his belief that he could actually achieve such a state of mind and freeze his internal world forever.

We could say, therefore, that the "circular ruins" were caused by the compelling unconscious choice of narcissistic omnipotence as against awareness of psychic reality. It is in the vicissitudes of the preservation or renunciation of omnipotence where the narcissistic dilemma of man confronted by time lies (Boschan, 1990). Mr A denied any awareness of passage of time in between sessions by diving into day-dreaming even before he closed the door, and perhaps only partially emerging from it when he arrived to his next session. In Borges's story, there is only a brief moment of conflict when the Old Man could have stopped dreaming but he did not appear to have had the possibility of making a choice given that he was already the product of a (his) dream. However, at the end, there is also some awareness of psychic conflict, and I would like to stress the words used by Borges to describe the Old Man's feelings when he realizes that he himself is also a dream and therefore cannot die: relief, humiliation, and terror. These three words brilliantly describe the conflict we all sustain between the wish to submerge ourselves in an idealised painless universe, and the horror that this possibility brings out in us. The terror, I think, expresses the Old Man's fear at the recognition of his self-destructiveness, of having given up on his human condition.

In everyday life Mr A's own needs, emotional and physical vulnerability, dependency, and fears are denied. He feels taken over by his omnipotent phantasies expressed through day-dreams, which he can recognize as the only things that give him pleasure and a sense of well being. Mr A lives, as Borges says, "lost in a maze of indefatigable mirrors". He became the dreamed son of another man's dream. But this other man is not the real father but the father he creates for himself, an idealized, omnipotent and all-powerful immortal God into whom he projects himself. For most of the time, the real parents, as well as the real him, become non-existent figures. Mr A has come to believe he only exists while he day-dreams and if he stops day-dreaming of his immortality he would die—he is his own Red King, in a fashion similar to what Sodre calls "death by day-dreaming" (1999). Winnicott (1971) described day-dreaming as an isolated phenomena that can remain static over the whole of the patient's life, absorbing energy but not contributing either to dreaming or to living, and very different to an imaginative process that is life-enriching. It seems to me that it is this "static quality" that is compulsively searched for, as it provides confirmation that it is possible to cancel time, change and differences.

The dream and the search to recover mortality

In Mr A's dream, the narcissistic withdrawal into an immortal Godlike him, who survives through being inside a drunken mother-vessel, seems to have been triggered by a feeling of exclusion and hatred of the father who actually throws him out of the boat; conveying through the ship and sea imagery how it must have felt to be the child of such a crew, of parents who, in their drunkenness, throw him out with only his omnipotence to save him. In the same way that he cancels all feelings at the end of his sessions and switches on his heroic day-dreams, he now submerges himself into the idealized alcoholic maternal waters, though there is some awareness of the restricted "bucket" into which he puts his mind. When he attempts to come out, what faces him is a couple of possibly adrift parents and a dead child. This dead child is the one who lacks a real identity, who has been drowned, whose humanity died like Homer's in "The city of immortals".

The anguish evoked in Mr A by the awareness of the dead baby-him, of the suicidal consequences of his identification with God, does for Mr A what the rain does for Homer. It is the awareness of this specular him that brings him back to the psychic reality of his attacks on life. In Mr A's dream the anguish encountered when seeing the dead child, held by this adrift couple, makes him desperately search for a woman. But, following his associations, this woman represented by Louise is an idealized woman—she is somebody who is associated with bringing back the lost illusion of re-establishing his balance when he fears a narcissistic loss. Mr A's sense of panic in the dream also stems from the feeling that he cannot resort to idealization as effectively as before when confronted with an internalized parental couple struggling for its psychic survival. However, at the time he brought this dream, I could also perceive a gradual movement in Mr A towards psychic change. There is some awareness of the restricted world of day-dreaming and a representation of the possibility of there being parents who could save their child, an implication that in his internal world such parents could be brought to life, probably through the experience in the transference of my concern for this dying child-him. Louise also represents me, and his wish for me to help him. I think that bringing this dream is his way of expressing his horror at what he is doing to himself and his objects, as well as his wish for recovery.

In contrast to the deadly circular self-contained world of day-dreaming, Mr A's dream could be seen as an expression of a desire to promote thinking, enquiry, and conjecture, of reaching a listener who is a separate object and not just the result of his own projections. Crying makes Mr A human again; it brings the lack, the difference, the absence. It also symbolizes his wish and hope to be helped by me not to go on killing himself. It opens the way to the process of mourning, to the possibility of coming alive without having to follow his parents and kill himself.

Mr A's dream conveys the powerful unconscious struggle he was going through, unlike his need to omnipotently control reality, in an "effort to get away from inner reality" (Winnicott, 1935, p. 130) via his repetitive, mindless omnipotent day-dreams. The discrepancy between his day-dreaming and this dream is similar to the difference noted by Britton between escapist and serious fiction. According to Britton, escapist fiction resembles obvious day-dreaming, while serious fiction expresses psychic reality, resonating with something unconscious and profoundly evocative (1995, 1998). Mr A's dream is a creative enterprise that stands in relation to his day-dreams, as Borges's stories stand in relation to his characters' dreams; in that both Mr A's dream and Borges's stories represent a desire to be in contact with internal reality.

These two stories by Borges amply resonate with our own unconscious wishes and beliefs, bringing together the in-depth unconscious workings of the dreamer and the aesthetic capacity to create thoughtfulness, mystery, and a sense of openness to the unexpected. But, as in "The circular ruins", the unconscious drama may well be compulsively repeated all over again and the inside of the day-dream bucket could be reinvested with life-giving qualities when it is a near suicidal place to live in.

I see the story "The circular ruins" as equivalent to Rufus's search for the City of Immortals. In "The circular ruins" the Old Man is trapped in the ruins that his narcissistic endeavour brings about. He cannot come alive. Though he has a conflict about that and he is terrified of his discovery, he is also relieved that the fire does not destroy him. He is relieved that he himself is a dream. In "The immortal" we can see Rufus's satisfaction at the possibility of being Homer and composing the Odyssey ("at least once"), of being a god, a hero, a philosopher, a demon, of being the world; it feels "divine" but also "terrible" and "incomprehensible". There seems to be some awareness that the destruction of his real identity and the creation of a world in which "all undertakings are in vain" is worse than real death.

It is only when Rufus sees what I think is his specular image of quasi-dead Homer that he recovers a sense of himself. He also recovers the wish to repair what he feels he has done to his internal objects (he returns the goldfish and hopes for a river to come and rescue him). Argos has torrents of tears running down his face and finally mutters, "Argos, Ulysses's dog". It is then that "being the dog of" (equivalent to being in relation to an object, such as being the brother or son of) can have some meaning. And, in the acceptance of the existence of a meaningful separate loved object, death becomes a necessary reality that can be mourned, allowing for language and identity to be re-established —as it is through language that our finitude is most radically revealed (Dastur, 1996). I think it is significant that in the Odyssey (Homer, 1980, p. 209), Argos is described as the dog brought up by Odysseus before he left. Argos waited for twenty years, lying unwanted on a pile of manure, only dying after seeing his old master again.

It is only now that Rufus can leave the mad gods. He gives up the search for immortality and struggles to recover his mortality, to own a specific, particular destiny, different to that of others and specific to himself. Awareness of transience, as was described by Freud (1916a), increases the value of life. It makes Rufus recover his peace of mind: "That night, I slept until dawn". In a paper on the fear of death, Segal describes that preparation for death involves a relinquishing of omnipotent control and the need to allow the objects to live on without one (1981b). The realization of the attacks on the good objects and the wish to repair the damage done to them give rise to a sense of loss and guilt. But this guilt is one that can now lead to reparation and to a way out from the circular ruins. Segal links this to the wish to "restore and re-create the lost loved object outside and within the ego", which she sees as the basis for sublimation and creativity (1981a).

Rufus is now able to abandon his compulsive quest for immortality, to find the 'spring of clear water" that makes him become aware of pain and finitude, of the differences between himself and others, of real death and the wish to live forever. He recovers his identity and is now able to write his own story, and leave behind a manuscript for others to read, possibly in the hope of achieving literary immortality.

The following poem by Borges (1923) I think beautifully describes the subject of this paper:

> Inscription on any tomb
> Let not the rash marble risk
> garrulous breaches of oblivion's omnipotence,
> in many words recalling
> name, renown events, birthplace.
> All those glass jewels are best left in the dark.
> Let not the marble say what men do not.
> The essentials of the dead man's life—
> the trembling hope,
> the implacable miracle of pain, the wonder of sensual delight—
> will abide forever.
> Blindly the wilful soul asks for length of days
> when its survival is assured by the lives of others,
> when you yourself are the embodied continuance
> of those who did not live into your time
> and others will be (and are) your immortality on earth.

Notes

1. I am grateful to Mrs Ignes Sodre for her helpful comments on this paper. A first version of this paper was presented at the Congress of the International Psychoanalytical Association, Barcelona, July 1997.

References

Anzieu, D. (1971). Le corps et le code dans les contes de J. L. Borges. *Nouvelle Revue de Psychanalytique*, 3: 177–205.

Aulagnier, P. C. (1975). *La violence de l'interpretation: du pictogramme a l'enonce*. Paris: PUF. English edition: *The Violence of Interpretation*. London: New Library of Psychoanalysis, 1975.

Barnatan, M. R. (1995). *Borges: biografia total*. Madrid: Temas de Hoy.

Bell-Villada, G. (1981). *Borges and his Fiction: A Guide to his Mind and Art*. Chapel Hill, NC: University of North Carolina Press.

Borges, J. L. (1923). Inscripcion en cualquier sepulcro. In: *Fervor de Buenos Aires*, Obras Completas. Buenos Aires: Emece Editores. English edition: *Jorge Luis Borges: Selected Poems*. Harmondsworth: Penguin, 1972.

Borges, J. L. (1944a). Tlon, Uqbar, Orbis Tertius. In: *Ficciones* (pp. 431–443), Jorge Luis Borges: Obras Completas. Buenos Aires: Emece Editores. English edition: *Labyrinths* (pp. 27–42). Harmondsworth: Penguin.

Borges, J. L. (1944b). Las ruinas circulares. In: *Ficciones* (pp. 451–455), Jorge Luis Borges: Obras Completas. Buenos Aires: Emece Editores. English edition: The circular ruins, in *Labyrinths* (pp. 72–77). Harmondsworth: Penguin.

Borges, J. L. (1949). El inmortal. In: *El Aleph* (pp. 533–544), Jorge Luis Borges: Obras Completas. Buenos Aires: Emece Editores. English edition: The immortal, in *Labyrinths* (pp. 135–149). Harmondsworth: Penguin.

Boschan, P. J. (1990). Temporality and narcissism. *International Review of Psycho-Analysis, 17*: 337–49.

Brenman, E. (1985). Cruelty and narrow-mindedness. *International Journal of Psychoanalysis, 66*: 273.

Britton, R. (1995). Reality and unreality in phantasy and fiction. In: E. S. Person, P. Fonagy, & S. A. Figueira (Eds.), *On Freud's "Creative Writers and Day-dreaming"* (82–105). Newhaven, CT: Yale University Press.

Britton, R. (1998). Daydream, phantasy and fiction. In: *Belief and Imagination*. London: Routledge, New Library of Psychoanalysis.

Carroll, L. (1911). *Through the Looking-Glass*. London: Penguin Classics.

Dastur, F. (1996). *Death—An Essay on Finitude*. London: Athlone.

Freud, S. (1914c). On narcissism: an introduction. *S.E., 14*: 73–102. London: Hogarth.

Freud, S. (1915b). Thoughts for the times on war and death. *S.E., 14*: 275–302. London: Hogarth.

Freud, S. (1915e). The unconscious. *S.E., 14*: 161–215. London: Hogarth.

Freud, S. (1916a). On transience. *S.E., 14*: 305–307. London: Hogarth.

Freud, S. (1920g). *Beyond the Pleasure Principle*. *S.E., 18*. London: Hogarth.

Homer (1980). *The Odyssey*. Oxford: Oxford University Press.

Pontalis, J.-B. (1977). On psychic pain. In: *Frontiers in Psychoanalysis* (pp. 194–205). London: Hogarth, 1981.

Rosenfeld, H. (1971). A clinical approach to the psychoanalytic theory of the life and death instincts: an investigation into the aggressive aspects of narcissism. *International Journal of Psychoanalysis, 52*: 169–178.

Salas, H. (1994). *Borges—una biografia*. Buenos Aires: Editorial Planeta.

Segal, H. (1981a). A psychoanalytic approach to aesthetics. In: *The Work of Hanna Segal* (pp. 185–206). New York: Jason Aronson.

Segal, H. (1981b). Fear of death: notes on the analysis of an old man. In: *The Work of Hanna Segal* (pp. 173–182). New York: Jason Aronson.

Sodre, I. (1999). Death by daydreaming: Madame Bovary. In: D. Bell (Ed.), *Psychoanalysis and Culture* (pp. 48–63). London: Duckworth.

Winnicott, D. W. (1935). The manic defence. In: *Through Paediatrics to Psycho-Analysis* (pp. 129–144). London: Hogarth.

Winnicott, D. W. (1971). Dreaming, fantasying and living: a case-history describing a primary dissociation. In: *Playing and Reality* (pp. 26–37). London: Tavistock Publications.

Woscoboinik, J. (1991). *El secreto de Borges—indagacion psicoanalitica de su obra*. Buenos Aires: Grupo Editor Latino-americano.

Zilboorg, G. (1938). The sense of immortality. *Psychoanalytic Quarterly*, 7: 171–199.

INDEX